Zhengyan Lin
Zhidong Bai

Probability Inequalities

Zhengyan Lin
Zhidong Bai

Probability Inequalities

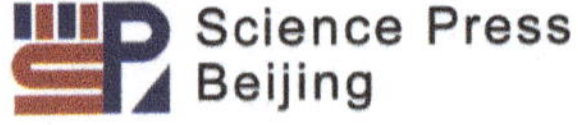

Printed on acid-free paper

Springer is part of Springer Science+Business Media (www.springer.com)

Authors
Zhengyan Lin
Department of Mathematics
Zhejiang University, China
zlin@zju.edu.cn

Zhidong Bai
School of Mathematics and Statistics
Northeast Normal University, China
baizd@nenu.edu.cn
Department of Statistics and Applied Probabilty
National University of Singapore, Singapore
stabaizd@leonis.nus.edu.sg

ISBN 978-7-03-025562-4
Science Press Beijing

ISBN 978-3-642-43777-9 ISBN 978-3-642-05261-3 (eBook)
Springer Heidelberg Dordrecht London New York

Softcover re-print of the Hardcover 1st edition 2011

Cover design: Frido Steinen-Broo, EStudio Calamar, Spain

Printed on acid-free paper

Springer is a part of Springer Science+Business Media (www.springer.com)

Preface

In almost every branch of quantitative sciences, inequalities play an important role in its development and are regarded to be even more important than equalities. This is indeed the case in probability and statistics. For example, the Chebyshev, Schwarz and Jensen inequalities are frequently used in probability theory, the Cramer-Rao inequality plays a fundamental role in mathematical statistics. Choosing or establishing an appropriate inequality is usually a key breakthrough in the solution of a problem, e.g. the Berry-Esseen inequality opens a way to evaluate the convergence rate of the normal approximation.

Research beginners usually face two difficulties when they start researching—they choose an appropriate inequality and/or cite an exact reference. In literature, almost no authors give references for frequently used inequalities, such as the Jensen inequality, Schwarz inequality, Fatou Lemma, etc. Another annoyance for beginners is that an inequality may have many different names and reference sources. For example, the Schwarz inequality is also called the Cauchy, Cauchy-Schwarz or Minkovski-Bnyakovski inequality. Bennet, Hoeffding and Bernstein inequalities have a very close relationship and format, and in literature some authors cross-cite in their use of the inequalities. This may be due to one author using an inequality and subsequent authors just simply copying the inequality's format and its reference without checking the original reference. All this may distress beginners very much.

The aim of this book is to help beginners with these problems. We provide a place to find the most frequently used inequalities, their proofs (if not too lengthy) and some references. Of course, for some of the more popularly known inequalities, such as Jensen and Schwarz, there is no necessity to give a reference and we will not do so.

The wording "frequently used" is based on our own understanding. It can be expected that many important probability inequalities are not collected in this work. Any comments and suggestions will be appreciated.

The writing of the book is supported partly by the National Science Foundation of China.

The authors would like to express their thanks to Ron Lim Beng Seng for improving our English in this book.

Zhengyan Lin

May, 2009

Contents

Chapter 1

Elementary Inequalities of Probabilities of Events

In this Chapter, we shall introduce some basic inequalities which can be found in many basic textbooks on probability theory, such as Feller (1968), Loéve (1977), etc.

We shall use the following popularly used notations. Let Ω be a space of elementary events, $\mathscr{F}$ be a σ-algebra of subsets of Ω, P be a probability measure defined on the events in $\mathscr{F}$. $(\Omega, \mathscr{F}, P)$ is so called a probability space. The events in $\mathscr{F}$ will be denoted by $A_1, A_2, \cdots$ or $A, B, \cdots$ etc. $A \bigcup B, AB$ (or $A \bigcap B$), $A - B$ and $A \Delta B$ denote the union, intersection, difference and symmetric difference of A and B respectively. A^c denotes the complement of A and $\varnothing$ denotes the empty set.

1.1 Inclusion-exclusion Formula

Let $A_1, A_2, \cdots, A_n$ be n events, then we have

$$\begin{aligned} P\left(\bigcup_{i=1}^n A_i\right) = & \sum_{i=1}^n P(A_i) - \sum_{1 \leqslant i < j \leqslant n} P(A_i A_j) + \cdots \\ & + (-1)^{k-1} \sum_{1 \leqslant i_1 < \cdots < i_k \leqslant n} P(A_{i_1} \cdots A_{i_k}) \\ & + \cdots + (-1)^{n-1} P(A_1 \cdots A_n). \end{aligned}$$

Proof. When $n = 2$, it is trivially known that

$$\begin{aligned} P\left(A_1 \bigcup A_2\right) = & P(A_1) + P(A_2 - A_1 A_2) \\ = & P(A_1) + P(A_2) - P(A_1 A_2). \end{aligned} \tag{1}$$

We show the formula by induction. Assume that the formula holds for n. We will show that it holds also for $n+1$. In fact, by (1) and the induction hypothesis,

$$P\left(\bigcup_{i=1}^{n+1} A_i\right) = P\left(\bigcup_{i=1}^{n} A_i\right) + P(A_{n+1}) - P\left(\bigcup_{i=1}^{n} A_i A_{n+1}\right)$$

$$= \sum_{i=1}^{n+1} P(A_i) - \sum_{1\leqslant i<j\leqslant n} P(A_iA_j) + \cdots + (-1)^{n-1}P(A_1\cdots A_n)$$

$$-\Bigg\{\sum_{i=1}^{n} P(A_iA_{n+1}) - \sum_{1\leqslant i<j\leqslant n} P(A_iA_jA_{n+1})$$

$$+ \cdots + (-1)^{n-1}P(A_1\cdots A_nA_{n+1})\Bigg\}$$

$$= \sum_{i=1}^{n+1} P(A_i) - \sum_{1\leqslant i<j\leqslant n+1} P(A_iA_j) + \cdots + (-1)^{n}P(A_1\cdots A_{n+1}).$$

1.2 Corollaries of the Inclusion-exclusion Formula

From the inclusion-exclusion formula, it is easy to deduce the following two conclusions.

1.2.a. When $A_1, \cdots, A_n$ are exchangeable, we have

$$P\left(\bigcup_{i=1}^{n} A_i\right) = \sum_{i=1}^{n} (-1)^{i-1}\binom{n}{i} P(A_1, \cdots, A_i).$$

Remark. A set of events $\{A_1, \cdots, A_n\}$ is said to be exchangeable if the probability of the intersection of any subset depends only on the size of the subset, that is, for any integers $1 \leqslant i_1 < \cdots < i_j \leqslant n$ and $1 \leqslant j \leqslant n, P(A_{i_1}A_{i_2}\cdots A_{i_j}) = p_j$.

1.2.b. When $A_1, \cdots, A_n$ are independent and $p = P(A_i)$, we have

$$p\left(\bigcup_{i=1}^{n} A_i\right) = \sum_{i=1}^{n} (-1)^{i-1}\binom{n}{i} p^i.$$

1.3 Further Consequences of the Inclusion-exclusion Formula

The following inequalities are also consequences of the inclusion-exclusion formula.

1.3.a. $\sum_{i=1}^{n} P(A_i) - \sum_{1\leqslant i<j\leqslant n} P(A_iA_j) \leqslant P\left(\bigcup_{i=1}^{n} A_i\right) \leqslant \sum_{i=1}^{n} P(A_i)$.

Remark. The right hand side (RHS) of the above inequality can be improved to

$$P\left(\bigcup_{i=1}^{n} A_i\right) \leqslant \sum_{i=1}^{n} P(A_i) - \sum_{i=2}^{n} P(A_1A_i).$$

Proof. When $n = 2$, the inequality with the improved RHS reduces to the inclusion-exclusion formula. Now, by induction we have

$$\begin{aligned} P\left(\bigcup_{i=1}^{n+1} A_i\right) &= P\left(\bigcup_{i=1}^{n} A_i\right) + P(A_{n+1}) - P\left(\left(\bigcup_{i=1}^{n} A_i\right)\bigcap A_{n+1}\right) \\ &\leqslant \sum_{i=1}^{n} P(A_i) - \sum_{i=2}^{n} P(A_1A_i) + P(A_{n+1}) - P(A_1A_{n+1}). \end{aligned}$$

This proves the improved right hand side. Similarly, by induction, we have

$$\begin{aligned} P\left(\bigcup_{i=1}^{n+1} A_i\right) &= P\left(\bigcup_{i=1}^{n} A_i\right) + P(A_{n+1}) - P\left(\left(\bigcup_{i=1}^{n} A_i\right)\bigcap A_{n+1}\right) \\ &\geqslant \sum_{i=1}^{n} P(A_i) - \sum_{i<j\leqslant n} P(A_iA_j) + P(A_{n+1}) - \sum_{i=1}^{n} P(A_iA_{n+1}). \end{aligned}$$

This proves the left hand side (LHS) of the inequality.

1.3.b. $|P(AB) - P(A)P(B)| \leqslant \frac{1}{4}$.

Proof.

$$\begin{aligned} |P(AB) - P(A)P(B)| &= |P(A)P(AB) + P(A^c)P(AB) \\ &\quad - P(A)P(AB) - P(A)P(A^cB)| \\ &= |P(A^c)P(AB) - P(A)P(A^cB)|. \end{aligned}$$

Since A^c and AB are disjoint, $P(A^c)P(AB) \leqslant 1/4$ (by noticing that $\max_{0<p<1} p(1-p) = 1/4$). Similarly, $P(A)P(A^cB) \leqslant 1/4$ as desired.

Remark. The difference $P(AB) - P(A)P(B)$ can be regarded as the covariance of the indicators I_A and I_B. The inequality 1.3.b can be easily proved by using the Cauchy-Schwarz inequality. Here, we proved the inequality by deliberately avoiding the use of moments.

1.3.c. $|P(A) - P(B)| \leqslant P(A\Delta B)$, $(A\Delta B = (A-B)\bigcup(B-A))$

Proof. By 1.3.a,

$$P(A\Delta B) \geqslant P(A-B) = P(A) - P(AB) \geqslant P(A) - P(B).$$

By the symmetry of A and B, 1.3.c is proved.

1.3.d. (Boole inequality). $P(AB) \geqslant 1 - P(A^c) - P(B^c)$.

Proof. $P(AB) + P(B^c) \geqslant P(A) = 1 - P(A^c)$.

1.3.e. Let $\limsup\limits_{n\to\infty} A_n = \bigcap\limits_{N=1}^{\infty}\bigcup\limits_{n=N}^{\infty} A_n$, $\liminf\limits_{n\to\infty} A_n = \bigcup\limits_{N=1}^{\infty}\bigcap\limits_{n=N}^{\infty} A_n$. Then

$$\begin{aligned} P(\liminf_{n\to\infty} A_n) &\leqslant \liminf_{n\to\infty} P(A_n) \leqslant \limsup_{n\to\infty} P(A_n) \\ &\leqslant P(\limsup_{n\to\infty} A_n) \leqslant \lim_{N\to\infty}\sum_{n=N}^{\infty} P(A_n). \end{aligned}$$

Proof. For any positive integer N, we have

$$\bigcap_{n=N}^{\infty} A_n \subset A_N \subset \bigcup_{n=N}^{\infty} A_n,$$

which simply implies

$$P\left(\bigcap_{n=N}^{\infty} A_n\right) \leqslant P(A_N) \leqslant P\left(\bigcup_{n=N}^{\infty} A_n\right) \leqslant \sum_{n=N}^{\infty} P(A_n).$$

Letting $N \to \infty$, we obtain the desired inequalities.

1.3.f. If $P(A) \geqslant 1-\varepsilon, P(B) \geqslant 1-\varepsilon$ for some $0 < \varepsilon < \frac{1}{2}$, then $P(AB) \geqslant 1 - 2\varepsilon$.

Proof. $P(AB) = P(A) + P(B) - P(A\bigcup B) \geqslant 1 - 2\varepsilon$.

1.3.g (Bonferroni inequality). Let $P_{[m]}(P_m)$ be the probability that exactly (at least, correspondingly) m events among $A_1, \cdots, A_n$ occur simultaneously. Putting

$$S_m = \sum_{1\leqslant i_1<\cdots<i_m\leqslant n} P(A_{i_1}\cdots A_{i_m}).$$

Then

$$S_m - (m+1)S_{m+1} \leqslant P_{[m]} \leqslant S_m, \qquad S_m - mS_{m+1} \leqslant P_m \leqslant S_m.$$

Proof. Let $A_{[m]}$ ($A_{(m)}$) denote the event that exactly (at least, correspondingly) m events among $A_1, \cdots, A_n$ happen simultaneously. Then we have

$$A_{[m]} \subset A_{(m)} = \bigcup_{1\leqslant i_1<\cdots<i_m\leqslant n} A_{i_1}\cdots A_{i_m}.$$

The RHS of the above inequalities follows by using the semi-additivity of probability measure.

On the other hand, we have

$$A_{[m]} \supset \bigcup_{1\leqslant i_1<\cdots<i_m\leqslant n} A_{i_1}\cdots A_{i_m} - \bigcup_{1\leqslant i_1<\cdots<i_{m+1}\leqslant n} A_{i_1}\cdots A_{i_{m+1}}.$$

This implies $p_{[m]} \geqslant p_m - S_{m+1}$. Also, for each $i_1 < \cdots < i_{m+1}$, the probability $P(A_{i_1}\cdots A_{i_{m+1}})$ is included at most in each of $P(A_{i'_1}\cdots A_{i'_m})$ in S_m, where $(i'_1,\cdots,i'_m)$ is a subset of $(i_1,\cdots,i_{m+1})$. Among the $m+1$, one needs to contribute to p_m. Therefore,

$$S_m - p_m \geqslant mS_{m+1}.$$

The LHS of 1.3.g then follow.

Remark. In fact, the inequalities 1.3.g can be proved from the following identities:

$$P_{[m]} = S_m - \binom{m+1}{m}S_{m+1} + \binom{m+2}{m}S_{m+2} + \cdots + (-1)^{n-m}\binom{n}{m}S_n,$$

$$P_m = S_m - \binom{m}{m-1}S_{m+1} + \binom{m+1}{m-1}S_{m+2} + \cdots + (-1)^{n-m}\binom{n-1}{m-1}S_n,$$

$$S_m = \sum_{i=m}^{n}\binom{i}{m}P_{[i]} \quad \text{and} \quad S_m = \sum_{i=m}^{n}\binom{i-1}{m-1}P_i.$$

By definition, we have

$$P_{[i]} = \sum_{F\in\mathscr{F}_i} P\Big(\bigcap_{j\in F} A_j \bigcap_{\ell\in F^c} A_\ell^c\Big),$$

where $\mathscr{F}_i$ is the collection of all subsets of size i of the set $\{1,2,\cdots,n\}$. Note that for each $\tilde{F}\in\mathscr{F}_m$, the set $\bigcap_{t\in\tilde{F}} A_t$ can be written as the union of disjoint subsets $\bigcap_{j\in F} A_j \bigcap_{\ell\in F^c} A_\ell^c$, for all $i \geqslant m$ where $F\in\mathscr{F}_i$ and $F\subset\tilde{F}$. This proves that

$$S_m = \sum_{i=m}^{n}\binom{i}{m}P_{[i]},$$

which implies $S_m = \sum_{i=m}^{n}\binom{i-1}{m-1}P_i$ by noticing that $P_{[i]} = P_i - P_{i+1}$.

Substituting the expression of S_m in terms of $P_{[i]}$ into the RHS of the first identity, we obtain

$$S_m - \binom{m+1}{m}S_{m+1} + \binom{m+2}{m}S_{m+2} + \cdots + (-1)^{n-m}\binom{n}{m}S_n$$

$$= \sum_{j=m}^{n}(-1)^{j-m}\binom{j}{m}\sum_{i=j}^{n}\binom{i}{j}P_{[i]}$$

$$= \sum_{i=m}^{n}\binom{i}{m}P_{[i]}\sum_{j=m}^{i}\binom{i-m}{j-m}(-1)^{j-m} = P_{[m]}.$$

By the same approach, one can prove the second identity.

1.4 Inequalities Related to Symmetric Difference

1.4.a. $P\left\{\left(\bigcup_n A_n\right)\Delta\left(\bigcup_n B_n\right)\right\} \leqslant P\left\{\bigcup_n\left(A_n\Delta B_n\right)\right\} \leqslant \sum_n P(A_n\Delta B_n).$

Proof. The left inequality follows from $\left(\bigcup_n A_n\right)\Delta\left(\bigcup_n B_n\right) \subset \bigcup_n (A_n\Delta B_n)$ by the definition of the symmetric difference. The right one follows from 1.3.a.

1.4.b. $P\{(A_1 - A_2)\Delta(B_1 - B_2)\} \leqslant P(A_1\Delta B_1) + P(A_2\Delta B_2).$

Proof. The inequality follows from the observation

$$(A_1 - A_2)\Delta(B_1 - B_2) \subset (A_1\Delta B_1)\bigcup(A_2\Delta B_2).$$

1.5 Inequalities Related to Independent Events

1.5.a. Let $\{A_n\}$ be a sequence of mutually independent events. Then

$$1 - P\left(\bigcup_{k=1}^{n} A_k\right) \leqslant \exp\left\{-\sum_{k=1}^{n} P(A_k)\right\},$$

$$1 - P\left(\bigcup_{k=1}^{\infty} A_k\right) \leqslant \lim_{n\to\infty}\exp\left\{-\sum_{k=1}^{n} P(A_k)\right\}.$$

Remark. The inequalities are useful in the proof of the Borel-Cantelli lemma.

Proof. The conclusions follow from an application of the inequality $1 - x \leqslant \mathrm{e}^{-x}$ for real x to the RHS of the identity

$$1 - P\left(\bigcup_{k=1}^{n} A_k\right) = P\left(\bigcap_{k=1}^{n} A_k^c\right) = \prod_{k=1}^{n}(1 - P(A_k)).$$

1.5.b. Suppose that A and B are independent, $AB \subset D$ and $A^cB^c \subset D^c$. Then $P(AD) \geqslant P(A)P(D)$.

Proof.

$$\begin{aligned} P(AD) &= P(ADB) + P(ADB^c) = P(AB) + P(AB^c) - P(AD^cB^c) \\ &= P(A)P(B) + P(AB^c) - P(D^cB^c) + P(A^cD^cB^c) \\ &= P(A)P(B) + P(AB^c) - P(D^cB^c) + P(A^cB^c) \\ &= P(A)P(B) + P(B^c) - P(D^cB^c) \\ &\geqslant P(A)P(BD) + P(A)P(B^cD) = P(A)P(D). \end{aligned}$$

1.5.c (Feller-Chung). Let $A_0 = \varnothing, \{A_n\}$ and $\{B_n\}$ be two sequences of events. Suppose that either

(i) B_n is independent of $A_nA_{n-1}^c \cdots A_0^c$ for all $n \geqslant 1$, or

(ii) B_n is independent of $\{A_n, A_nA_{n+1}^c, A_nA_{n+1}^cA_{n+2}^c, \cdots\}$ for all $n \geqslant 1$. Then

$$P\left(\bigcup_{n=1}^{\infty} A_nB_n\right) \geqslant \inf_{n\geqslant 1} P(B_n)P\left(\bigcup_{n=1}^{\infty} A_n\right).$$

Proof. In case (i),

$$\begin{aligned} P\left\{\bigcup_{n=1}^{\infty} A_nB_n\right\} &= P\left\{\bigcup_{n=1}^{\infty} B_nA_n \bigcap_{j=0}^{n-1} (B_jA_j)^c\right\} = \sum_{n=1}^{\infty} P\left\{B_nA_n \bigcap_{j=0}^{n-1} (B_jA_j)^c\right\} \\ &\geqslant \sum_{n=1}^{\infty} P\left\{B_nA_n \bigcap_{j=0}^{n-1} A_j^c\right\} = \sum_{n=1}^{\infty} P(B_n)P\left\{A_n \bigcap_{j=0}^{n-1} A_j^c\right\} \\ &\geqslant \inf_{n\geqslant 1} P(B_n)P\left(\bigcup_{n=1}^{\infty} A_n\right), \end{aligned}$$

and in case (ii),

$$\begin{aligned} P\left\{\bigcup_{j=1}^{n} A_jB_j\right\} &= \sum_{j=1}^{n} P\left\{A_jB_j \bigcap_{i=j+1}^{n} (A_iB_i)^c\right\} \geqslant \sum_{j=1}^{n} P\left\{A_jB_j \bigcap_{i=j+1}^{n} A_i^c\right\} \\ &= \sum_{j=1}^{n} P(B_j)P\left\{A_j \bigcap_{i=j+1}^{n} A_i^c\right\} \geqslant \inf_{1\leqslant j\leqslant n} P(B_j)P\left\{\bigcup_{j=1}^{n} A_j\right\}. \end{aligned}$$

1.6 Lower Bound for Union (Chung-Erdös)

$$P\left(\bigcup_{i=1}^{n} A_i\right) \geqslant \left(\sum_{i=1}^{n} P(A_i)\right)^2 \Big/ \left(\sum_{i=1}^{n} P(A_i) + 2\sum_{1\leqslant i<j\leqslant n} P(A_iA_j)\right).$$

Proof. Define random variables $X_k(\omega), \omega \in \Omega$, by

$$X_i(\omega) = \begin{cases} 0, & \text{if } \omega \notin A_i, \\ 1, & \text{if } \omega \in A_i. \end{cases}$$

Then

$$2\sum_{1\leqslant i<j\leqslant n} P(A_iA_j) = E(X_1 + \cdots + X_n)^2 - E(X_1^2 + \cdots + X_n^2).$$

By the Cauchy-Schwarz inequality (see 8.4.b), we have

$$(E(X_1 + \cdots + X_n))^2 \leqslant P(X_1 + \cdots + X_n > 0)E(X_1 + \cdots + X_n)^2.$$

Note that $EX_i = EX_i^2 = P(A_i), P(X_1 + \cdots + X_n > 0) = P\left(\bigcup_{i=1}^{n} A_i\right)$ by definition. Combining the above two relations yields the desired inequality.

References

Chow Y S, Teicher H. 1997. Probability Theory: Independence, Interchangeability, Martingales. 2nd ed.. New York: Springer-Verlag.

Chung K L. 1974. A Course in Probability Theory. 2nd ed.. New York: Academic Press.

Feller W. 1968. An Introduction to Probability Theory and Its Applications. 3rd ed.. New York: Wiley.

Hardy G H, Littlewood J E, Polya G. 1952. Inequalities. Cambridge: Cambridge University Press.

Loéve M. 1977. Probability Theory. 4th ed.. New York: Springer-Verlag.

Petrov V V. 1995. Limit Theorems of Probability Theory: Sequences of Independent Random Variables. Oxford & New York: Clarendon Press & Oxford University Press.

Shorack G R, Wellner J A. 1986. Empirical Processes with Applications to Statistics. New York: Wiley.

Chapter 2

Inequalities Related to Commonly Used Distributions

Commonly used distributions play an important role in applied statistics, statistical computing and applied probability. So, inequalities related to these distributions are of great interest in these areas.

Let ξ be a random variable (r.v.). Then its distribution function (d.f.) is defined by $F(x) = P(\xi < x)$ and its probability density function (pdf.) $p(x)$ (if it exists) is defined to be a measurable function such that $F(x) = \int_{-\infty}^{x} p(y)\mathrm{d}y$. Write

$$\Phi(x) = \frac{1}{\sqrt{2\pi}} \int_{-\infty}^{x} \mathrm{e}^{-t^2/2}\mathrm{d}t \qquad \text{and} \qquad \varphi(x) = \frac{1}{\sqrt{2\pi}}\mathrm{e}^{-x^2/2}$$

for the standard normal d.f. and pdf. respectively,

$$b(k; n, p) = \binom{n}{k} p^k q^{n-k}, \quad k = 0, 1, \cdots, n, \quad 0 < p < 1, \quad q = 1 - p$$

for the binomial distribution with parameters n and p,

$$p(k; \lambda) = \frac{\lambda^k \mathrm{e}^{-\lambda}}{k!}, \qquad k = 0, 1, \cdots, \quad \lambda > 0$$

for the Poisson distribution with parameter λ.

2.1 Inequalities Related to the Normal d.f.

2.1.a. $\frac{1}{\sqrt{2\pi}}(b-a)\exp\{-(a^2 \vee b^2)/2\} \leqslant \Phi(b) - \Phi(a) \leqslant \frac{1}{\sqrt{2\pi}}(b-a), -\infty < a < b < \infty$.

Proof. It follows from the fact that $\mathrm{e}^{-x^2/2}$ on $[a, b]$ is between $\exp\{-(a^2 \vee b^2)/2\}$ and 1.

2.1.b. For all $x > 0$,

$$\left(\frac{1}{x} - \frac{1}{x^3}\right)\varphi(x) < \frac{x}{1+x^2}\varphi(x) < 1 - \Phi(x) < \frac{1}{x}\varphi(x).$$

Proof. The most right inequality follows from integration by parts

$$\int_x^\infty \mathrm{e}^{-t^2/2}\mathrm{d}t = \frac{1}{x}\mathrm{e}^{-x^2/2} - \int_x^\infty \frac{1}{t^2}\mathrm{e}^{-t^2/2}\mathrm{d}t$$

for all $x > 0$. The most left inequality is elementary. The middle inequality follows from the observation that for all $x > 0$,

$$\frac{1}{x^2}\int_x^\infty \mathrm{e}^{-t^2/2}\mathrm{d}t > \int_x^\infty \frac{1}{t^2}\mathrm{e}^{-t^2/2}\mathrm{d}t = \frac{1}{x}\mathrm{e}^{-x^2/2} - \int_x^\infty \mathrm{e}^{-t^2/2}\mathrm{d}t.$$

Hence

$$\frac{1}{x}\mathrm{e}^{-x^2/2} < \left(1 + \frac{1}{x^2}\right)\int_x^\infty \mathrm{e}^{-t^2/2}\mathrm{d}t,$$

which implies what is to be proven.

Remark. By repeatedly integrating by parts, the above inequality can be extended as, for any integer $k \geqslant 0$ and $x > 0$,

$$\sum_{j=0}^{2k+1} \frac{(-1)^j(2j-1)!!}{x^{2j+1}}\varphi(x) < 1 - \Phi(x) < \sum_{j=0}^{2k} \frac{(-1)^j(2j-1)!!}{x^{2j+1}}\varphi(x),$$

where $(2j-1)!! = (2j-1)\cdots 3\cdot 1 = \frac{(2j)!}{2^j j!}$ and by convention $(-1)!! = 1$. The reader is reminded that one can only do finite steps of the integration by parts because the series $\sum\limits_{j=0}^{\infty} \frac{(-1)^j(2j+1)!!}{x^{2j+1}}$ does not converge for any $x > 0$. That means, one cannot get an identity by making $k \to \infty$.

2.1.c. For all real x, $1 - \Phi(x) \geqslant \frac{1}{2}(\sqrt{x^2+4} - x)\varphi(x)$ for all $x > -1, 1 - \Phi(x) \leqslant \frac{4}{3x+\sqrt{x^2+8}}\varphi(x)$.

Proof. Using the Cauchy-Schwarz inequality (see 8.4.b), we have

$$\begin{aligned}(\mathrm{e}^{-x^2/2})^2 &= \left(\int_x^\infty t\mathrm{e}^{-t^2/2}\mathrm{d}t\right)^2 \leqslant \left(\int_x^\infty t^2\mathrm{e}^{-t^2/2}\mathrm{d}t\right)\left(\int_x^\infty \mathrm{e}^{-t^2/2}\mathrm{d}t\right) \\ &= \left(x\mathrm{e}^{-x^2/2} + \int_x^\infty \mathrm{e}^{-t^2/2}\mathrm{d}t\right)\int_x^\infty \mathrm{e}^{-t^2/2}\mathrm{d}t,\end{aligned}$$

which implies the first inequality. Let

$$\nu_x = \mathrm{e}^{-x^2/2}\Big/\int_x^\infty \mathrm{e}^{-t^2/2}\mathrm{d}t \quad \text{and} \quad \varphi_x = (\nu_x - x)(2\nu_x - x).$$

By the LHS of 2.1.b, for all $x > 0$, we have $\nu_x > \frac{x^3}{x^2-1}\varphi(x)$ and hence for all $x > 0$,

$$\varphi_x > \frac{x^2(x^2+1)}{(x^2-1)^2} > 1.$$

Next we show that $\varphi_x > 1$ is true for all x. If not, by continuity, there is an x_0 such that

$$\varphi_{x0} = 1, \qquad \varphi'_{x0} \geqslant 0.$$

But

$$\begin{aligned}\varphi'_{x0} &= (\nu'_{x0} - 1)(2\nu_{x0} - x_0) + (\nu_{x0} - x_0)(2\nu'_{x0} - 1)\\ &= \nu_{x0}(\varphi_{x0} - 1) + 2(\nu_{x0} - x_0)(\nu'_{x0} - 1)\\ &= 2(\nu_{x0} - x_0)(\nu'_{x0} - 1).\end{aligned}$$

By the RHS of 2.1.b, we have $\nu_x - x > 0$ for all real x. By the assumption $1 = \varphi_{x0} = 2\nu^2_{x0} - 3x_0\nu_{x0} + x_0^2$, we have

$$\nu'_{x0} - 1 = \nu^2_{x0} - x_0\nu_{x0} - [2\nu^2_{x0} - 3x_0\nu_{x0} + x_0^2] = -(x_0 - \nu_{x0})^2 < 0,$$

which implies that $\varphi'_{x0} < 0$, contradicting the assumption that $\varphi'_{x0} \geqslant 0$. Hence, for finite x,

$$\varphi_x > 1.$$

Considering the above inequality as a quadratic inequality in ν_x, we obtain that for all x, either

$$\nu_x > \frac{3x + \sqrt{x^2+8}}{4} \quad \text{or} \quad \nu_x < \frac{3x - \sqrt{x^2+8}}{4}.$$

It is obvious that the first inequality is true for $x \leqslant 0$ since the second one is impossible. By continuity, we conclude that the first inequality is true for all real x. Then, the second inequality of 2.1.c follows from the fact that $\frac{3x+\sqrt{x^2+8}}{4} > 0$ for all $x > -1$.

2.1.d. $1-\Phi(x) \sim \varphi(x)\{\frac{1}{x} - \frac{1}{x^3} + \frac{1\cdot 3}{x^5} - \cdots + (-1)^k\frac{(2k-1)!!}{x^{2k+1}}\}$ as $x \to \infty$ and for $x > 0$ the RHS overestimates $1-\Phi(x)$ if k is even, and underestimates if k is odd.

Proof. See the remark to 2.1.b.

2.1.e. Let (X, Y) be a bivariate normal random vector with the distribution

$$N\left(\begin{pmatrix}0\\0\end{pmatrix}, \begin{pmatrix}1 & r\\ r & 1\end{pmatrix}\right).$$

If $0 \leqslant r < 1$, then for any real a and b,

$$(1-\Phi(a))\left(1-\Phi\left(\frac{b-ra}{\sqrt{1-r^2}}\right)\right) \leqslant P(X>a, Y>b)$$

$$\leqslant (1-\Phi(a))\left\{\left(1-\Phi\left(\frac{b-ra}{\sqrt{1-r^2}}\right)\right)+r\frac{\varphi(b)}{\varphi(a)}\left(1-\Phi\left(\frac{a-rb}{\sqrt{1-r^2}}\right)\right)\right\}.$$

If $-1 < r \leqslant 0$, the inequalities are reversed.

Proof. By integration by parts we get

$$\begin{aligned} P(X>a,Y>b) &= \int_a^\infty \varphi(x)\left(1-\Phi\left(\frac{b-rx}{\sqrt{1-r^2}}\right)\right)\mathrm{d}x \\ &= (1-\Phi(a))\left(1-\Phi\left(\frac{b-ra}{\sqrt{1-r^2}}\right)\right) \\ &\quad + \int_a^\infty (1-\Phi(x))\varphi\left(\frac{b-rx}{\sqrt{1-r^2}}\right)\frac{r}{\sqrt{1-r^2}}\mathrm{d}x. \end{aligned}$$

Suppose $0 \leqslant r < 1$. The lower bound then follows immediately. Next note that $(1-\Phi(x))/\varphi(x)$ is decreasing. Thus

$$\begin{aligned} &\int_a^\infty (1-\Phi(x))\varphi\left(\frac{b-rx}{\sqrt{1-r^2}}\right)\frac{\mathrm{d}x}{\sqrt{1-r^2}} \\ \leqslant &\frac{1-\Phi(a)}{\varphi(a)}\int_a^\infty \varphi(x)\varphi\left(\frac{b-rx}{\sqrt{1-r^2}}\right)\frac{\mathrm{d}x}{\sqrt{1-r^2}} \\ = &\frac{1-\Phi(a)}{\varphi(a)}\int_a^\infty \varphi(b)\varphi\left(\frac{x-rb}{\sqrt{1-r^2}}\right)\frac{\mathrm{d}x}{\sqrt{1-r^2}} \\ = &(1-\Phi(a))\frac{\varphi(b)}{\varphi(a)}\left(1-\Phi\left(\frac{a-rb}{\sqrt{1-r^2}}\right)\right), \end{aligned}$$

which gives the upper bound. For the case $-1 < r \leqslant 0$ the same argument works by noting the reversed directions of the inequalities.

2.2 Slepian Type Inequalities

2.2.a (Slepian lemma). Let $(X_1, \cdots, X_n)$ be a normal random vector with $EX_j = 0$, $EX_j^2 = 1$, $j = 1, \cdots, n$. Put $\gamma_{kl} = EX_kX_l$ and $\Gamma = (\gamma_{kl})$ which is the covariance matrix. Let $I_x^{+1} = [x, \infty), I_x^{-1} = (-\infty, x)$ and $A_j = \{X_j \in I_{x_j}^{\varepsilon_j}\}$, where ε_j is either $+1$ or -1. Then $P\left\{\bigcap_{j=1}^n A_j; \Gamma\right\}$ is an increasing function of γ_{kl} if $\varepsilon_k\varepsilon_l = 1$; otherwise it is decreasing.

Proof. The pdf of $(X_1, \cdots, X_n)$ can be written in terms of its characteristic function (c.f.) by[1]

$$p(x_1, \cdots, x_n; \Gamma) = (2\pi)^{-n} \int \cdots \int \exp\left\{ \mathrm{i} \sum_{j=1}^{n} t_j x_j - \frac{1}{2} \sum_{k,l} \gamma_{kl} t_k t_l \right\} \mathrm{d}t_1 \cdots \mathrm{d}t_n.$$

It follows that

$$\frac{\partial p}{\partial \gamma_{kl}} = \frac{\partial^2 p}{\partial x_k \partial x_l}, \quad 1 \leqslant k < l \leqslant n.$$

Hence we regard p as a function of the $n(n-1)/2$ variables $\gamma_{kl}, k < l$. Moreover

$$P\left\{ \bigcap_{j=1}^{n} A_j; \Gamma \right\} = \int_{I_{x_1}^{\varepsilon_1}} \cdots \int_{I_{xn}^{\varepsilon n}} p(u_1, \cdots, u_n; \Gamma) \mathrm{d}u_1 \cdots \mathrm{d}u_n.$$

Consider the probability as a function of $\gamma_{kl}, k < l$, we then prove the conclusion by verifying the nonnegativity of an example. Consider γ_{12} and the integral intervals are $I_{x_1}^{+1}$ and $I_{x_2}^{+1}$. We have

$$\begin{aligned} &\frac{\partial P\left\{ \bigcap_{j=1}^{n} A_j; \Gamma \right\}}{\partial \gamma_{12}} \\ &= \int_{I_{x_1}^{+1}} \int_{I_{x_2}^{+1}} \cdots \int_{I_{xn}^{\varepsilon n}} \frac{\partial^2}{\partial u_1 \partial u_2} p(u_1, u_2, \cdots, u_n; \Gamma) \mathrm{d}u_1 \mathrm{d}u_2 \cdots \mathrm{d}u_n \\ &= \int_{I_{x_3}^{\varepsilon_3}} \cdots \int_{I_{xn}^{\varepsilon n}} p(x_1, x_2, u_3, \cdots, u_n; \Gamma) \mathrm{d}u_3 \cdots \mathrm{d}u_n \geqslant 0. \end{aligned}$$

Hence $P\left\{ \bigcap_{j=1}^{n} A_j; \Gamma \right\}$ is an increasing function of γ_{12}.

2.2.b (Berman). Continue to use the notation in 2.2.a. We have

$$\left| P\left\{ \bigcap_{j=1}^{n} A_j \right\} - \prod_{j=1}^{n} P(A_j) \right| \leqslant \sum_{1 \leqslant k < l \leqslant n} |\gamma_{kl}| \varphi(x_k, x_l; \gamma_{kl}^*),$$

where $\varphi(x, y; \gamma_{kl}^*)$ is the standard bivariate normal density function with a covariance γ_{kl}^*, and γ_{kl}^* is a number between 0 and γ_{kl}.

[1] In this book, "i" sometimes denotes an index or at other times denotes the image unit. It is not confused from the context.

Proof. Put $I_j = I_{x_j}^{\varepsilon_j}$,

$$Q((I_1, \cdots, I_n); \Gamma) = \int_{I_1} \cdots \int_{I_n} p(u_1, \cdots, u_n; \Gamma) du_1 \cdots du_n$$

and I to be an identity matrix of order n. Then, by the mean value theorem, there exist numbers γ_{kl}^* between 0 and γ_{kl} such that

$$\begin{aligned} P\left\{\bigcap_{j=1}^n A_j; \Gamma\right\} - P\left\{\bigcap_{j=1}^n A_j; I\right\} &= \sum_{1 \leqslant k < l \leqslant n} \gamma_{kl} (\partial Q / \partial \gamma_{kl})((I_1, \cdots .I_n); (\gamma_{kl}^*)) \\ &\leqslant \sum_{1 \leqslant k < l \leqslant n} |\gamma_{kl}| \varphi(x_k, x_l; \gamma_{kl}^*). \end{aligned}$$

2.2.c (Gordon). Let $\{X_{ij}\}$ and $\{Y_{ij}\}$ $(1 \leqslant i \leqslant n, 1 \leqslant j \leqslant m)$ be two collections of normal r.v.'s satisfying:

(1) $EX_{ij} = EY_{ij} = 0, EX_{ij}^2 = EY_{ij}^2,\ 1 \leqslant i \leqslant n, 1 \leqslant j \leqslant m$;

(2) $E(X_{ij}X_{ik}) \leqslant E(Y_{ij}Y_{ik}),\ 1 \leqslant i \leqslant n, 1 \leqslant j, k \leqslant m$;

(3) $E(X_{ij}X_{lk}) \geqslant E(Y_{ij}Y_{lk}),\ i \neq l, 1 \leqslant i, l \leqslant n, 1 \leqslant j, k \leqslant m$.

Then for any x_{ij},

$$P\left\{\bigcap_{i=1}^n \bigcup_{j=1}^m (X_{ij} \geqslant x_{ij})\right\} \geqslant P\left\{\bigcap_{i=1}^n \bigcup_{j=1}^m (Y_{ij} \geqslant x_{ij})\right\}.$$

Proof. Denote a vector $x = (x_1, \cdots, x_{nm})$ in R^{nm} by

$$x = (x_{11}, \cdots, x_{1m}, x_{21}, \cdots, x_{2m}, \cdots, x_{n1}, \cdots, x_{nm}),$$

where $x_{ij} = x_{(i-1)m+j},\ 1 \leqslant i \leqslant n,\ 1 \leqslant j \leqslant m$.

For a given positive definite matrix $\Gamma = (\gamma_{uv})_{nm}$, let $Z = (Z_1, \cdots, Z_{nm})$ be a centered normal random vector with covariance matrix Γ. Then its pdf can be written as

$$g(Z; \Gamma) = (2\pi)^{-nm} \int_{R^{nm}} \exp\left\{\mathrm{i}(x, Z) - \frac{1}{2} x' \Gamma x\right\} dx.$$

As mentioned in subsection 2.2.a, if $u \neq v$ then $\partial g / \partial \gamma_{uv} = \partial^2 g / \partial z_u \partial z_v$. Notice that if $u = (i-1)m + j, v = (l-1)m + k \quad (1 \leqslant i, l \leqslant n, 1 \leqslant j, k \leqslant m)$, then by our notation

$$\gamma_{uv} = E(Z_u Z_v) = E(Z_{ij} Z_{lk}).$$

Let $A_{ij} = \{Z_{ij} \geqslant x_{ij}\}$, $B_{i0} = A_{i1}$ and $B_{ij} = A_{i1}^c \cdots A_{ij}^c A_{i,j+1}$. Then we can verify that

$$\bigcap_{i=1}^{n} \bigcup_{j=1}^{m} A_{ij} = \bigcup_{j_1=0}^{m-1} \cdots \bigcup_{j_n=0}^{m-1} (B_{1j_1} B_{2j_2} \cdots B_{nj_n}).$$

By this relation we obtain

$$Q(Z;\Gamma) \equiv P\left(\bigcap_{i=1}^{n} \bigcup_{j=1}^{m} A_{ij}\right) = \sum_{j_1=0}^{m-1} \cdots \sum_{j_n=0}^{m-1} \int_{B_{njn}} \cdots \int_{B_{1j_1}} g(z)\mathrm{d}z,$$

where we have used the facts that for any function $f(z_{11} \cdots z_{nm})$ and indices $1 \leqslant i \leqslant n$, $0 \leqslant j \leqslant m-1$,

$$\int_{B_{i0}} f(z)\mathrm{d}z_{i1} \cdots \mathrm{d}z_{im} = \int_{x_{i1}}^{\infty} \int_{-\infty}^{\infty} \cdots \int_{-\infty}^{\infty} f(z)\mathrm{d}z_{im} \cdots \mathrm{d}z_{i2}\mathrm{d}z_{i1}$$

and

$$\int_{B_{ij}} f(z)\mathrm{d}z_{i1} \cdots \mathrm{d}z_{im} = \int_{-\infty}^{x_{i1}} \cdots \int_{-\infty}^{x_{ij}} \int_{x_{i,j+1}}^{\infty} \int_{-\infty}^{\infty} \cdots$$
$$\int_{-\infty}^{\infty} f(z)\mathrm{d}z_{im} \cdots \mathrm{d}z_{i2}\mathrm{d}z_{i1}.$$

By differentiating Q with respect to γ_{uv} we obtain

$$\frac{\partial Q(z;\Gamma)}{\partial \gamma_{uv}} = \sum_{j_1=0}^{m-1} \cdots \sum_{j_n=0}^{m-1} \int_{B_{1j_1}} \cdots \int_{B_{njn}} \frac{\partial^2 g(z)}{\partial z_u \partial z_v}\mathrm{d}z, \qquad u \neq v.$$

There are two possibilities for the above integrals:

(a) $u = (i-1)m+k$, $v = (i-1)m+l$, where $1 \leqslant k < l \leqslant m$, $1 \leqslant i \leqslant n$;

(b) $u = (i-1)m+k$, $v = (i'-1)m+l$, where $1 \leqslant k < l \leqslant m$, $1 \leqslant i < i' \leqslant n$.

In case (a), without loss of generality, we take $z_u = z_{1,m-1}, z_v = z_{1m}$(i.e, $i=1, k=m-1, l=m$), then

$$\int_{B_{1j_1}} \frac{\partial^2 g(z)}{\partial z_{1,m-1}\partial z_{1m}}\mathrm{d}z_{11} \cdots \mathrm{d}z_{1m}$$
$$= \int_{-\infty}^{x_{11}} \cdots \int_{-\infty}^{x_{1j_1}} \int_{x_{1,j_1+1}}^{\infty} \int_{-\infty}^{\infty} \cdots \int_{-\infty}^{\infty} \frac{\partial^2 g(z)}{\partial z_{1,m-1}\partial z_{1m}}\mathrm{d}z_{1m} \cdots \mathrm{d}z_{11}$$

and we see that this is equal to zero if $j_1 < m-1$ because

$$\int_{-\infty}^{\infty} \frac{\partial^2 g(z)}{\partial z_{1,m-1}\partial z_{1m}} \mathrm{d}z_{1m} = 0.$$

But when $j_1 = m-1$, then

$$\int_{B_{1j_1}} \frac{\partial^2 g(z)}{\partial z_{1,m-1}\partial z_{1m}} \mathrm{d}z_{11}\cdots \mathrm{d}z_{1m}$$
$$= -\int_{-\infty}^{x_{11}} \cdots \int_{-\infty}^{x_{1,m-2}} g(z)|_{z_{1,m-1}=x_{1,m-1}, z_{1m}=x_{1m}} \mathrm{d}z_{1,m-2}\cdots \mathrm{d}z_{11}.$$

Hence, it follows that in case (a), $\partial Q/\partial \gamma_{uv} \leqslant 0$.

In case (b), without loss of generality, we take $z_u = z_{1m}$ and $z_v = z_{2m}$. Then, when either j_1 or j_2 is smaller than $m-1$, we obtain as in the above manner,

$$\int_{B_{2j_2}} \int_{B_{1j_1}} \frac{\partial^2 g(z)}{\partial z_{1m}\partial z_{2m}} \mathrm{d}z_{11}\cdots \mathrm{d}z_{1m}\mathrm{d}z_{21}\cdots \mathrm{d}z_{2m} = 0.$$

However, if $j_1 = j_2 = m-1$, then

$$\int_{B_{2,m-1}} \int_{B_{1,m-1}} \frac{\partial^2 g(z)}{\partial z_{1m}\partial z_{2m}} \mathrm{d}z_{11}\cdots \mathrm{d}z_{1m}\mathrm{d}z_{21}\cdots \mathrm{d}z_{2m}$$
$$= \int_{-\infty}^{x_{11}} \cdots \int_{-\infty}^{x_{1,m-1}} \int_{x_{1m}}^{\infty} \int_{-\infty}^{x_{21}} \cdots \int_{-\infty}^{x_{2,m-1}} \int_{x_{2m}}^{\infty} \frac{\partial^2 g(z)}{\partial z_{1m}\partial z_{2m}} \mathrm{d}z_{2m}\cdots \mathrm{d}z_{11}$$
$$= \int_{-\infty}^{x_{11}} \cdots \int_{-\infty}^{x_{1,m-1}} \int_{-\infty}^{x_{21}} \cdots \int_{-\infty}^{x_{2,m-1}} g(z)|_{z_{1m}=x_{1m}, z_{2m}=x_{2m}} \mathrm{d}z_{2,m-1}\cdots$$
$$\mathrm{d}z_{21}\mathrm{d}z_{1,m-1}\cdots \mathrm{d}z_{11}$$
$$\geqslant 0.$$

Hence it follows that in case (b), $\partial Q/\partial \gamma_{uv} \geqslant 0$.

Let Γ_X and Γ_Y be the covariance matrices of $X = (X_{11}, \cdots, X_{1m}, \cdots, X_{n1}, \cdots, X_{nm})$ and $Y = (Y_{11}, \cdots, Y_{1m}, \cdots, Y_{n1}, \cdots, Y_{nm})$. By a standard approximation procedure we may assume that Γ_X and Γ_Y are both positive definite. For $0 \leqslant \theta \leqslant 1$, let $\Gamma_X = (\gamma_{uv})$ and $\Gamma_Y = (s_{uv})$ and $\Gamma(\theta) = \theta\Gamma_X + (1-\theta)\Gamma_Y$. By assumption (1), $\gamma_{uv} = s_{uv}$ for all u, therefore

$$\frac{\mathrm{d}Q(z;\Gamma(\theta))}{\mathrm{d}\theta} = \sum_{u<v} \frac{\partial Q(z;\Gamma)}{\partial \gamma_{uv}}|_{\gamma=\gamma(\theta)}(\gamma_{uv} - s_{uv}).$$

By the assumptions (2) and (3), $\gamma_{uv} \leqslant s_{uv}$ in case (a) and $\gamma_{uv} \geqslant s_{uv}$ in case (b), hence $\mathrm{d}Q/\mathrm{d}\theta \geqslant 0$. Therefore $Q(z, \Gamma(1)) \geqslant Q(z, \Gamma(0))$, i.e., $Q(X; \Gamma_X) \geqslant Q(Y; \Gamma_Y)$. The proof is completed.

Remark. As a consequence, under the conditions (1), (2) and (3),

$$E \min_{1\leqslant i\leqslant n} \max_{1\leqslant j\leqslant m} X_{ij} \geqslant E \min_{1\leqslant i\leqslant n} \max_{1\leqslant j\leqslant m} Y_{ij}.$$

2.3 Anderson Type Inequalities

2.3.a. Let X be a zero-mean Gaussian vector in $\mathbb{R}^N$ and let D be a convex set in $\mathbb{R}^N$ symmetric with respect to the origin. Then for any $0 \leqslant |h| \leqslant 1$ we have

$$P\{X + x \in D\} \leqslant P\{X + hx \in D\}.$$

Proof. This is a direct consequence of the following integral inequalities (Anderson, 1955): For the convex set D of $\mathbb{R}^N$ where D is symmetric with respect to the origin, a non-negative function $f(x)$ in $\mathbb{R}^N$ satisfies:

(i) $f(x) = f(-x)$;

(ii) for any $u > 0$, $\{x : f(x) \geqslant u\}$ is a convex set;

(iii) $\int_D f(x)\,\mathrm{d}x < \infty$ (under the meaning of Lebesgue integral).

Then for any $0 \leqslant |h| \leqslant 1$, we have

$$\int_D f(x + y)\,\mathrm{d}x \leqslant \int_D f(x + hy)\,\mathrm{d}x. \tag{2}$$

Assume that $f(x) = (2\pi)^{-N/2} \exp\{-x'\Sigma^{-1}x/2\}$, in which Σ is the positive definite covariance matrix of X, the proof follows from (2).

2.3.b. Let X_1 and X_2 be Gaussian vectors with means zero and covariance matrixes Σ_1 and Σ_2, respectively. If $\Sigma_2 - \Sigma_1$ is positive semi-definite and D is a convex set symmetric with respect to the origin, then

$$P\{X_1 \in D\} \geqslant P\{X_2 \in D\}.$$

Proof. Let Y be a Gaussian vector in $\mathbb{R}^N$ with mean zero and covariance matrix $\Sigma_2 - \Sigma_1$, and is independent of X_1. Then X_2 and $X_1 + Y$ have identical distributions. From 2.3.a, we have

$$\begin{aligned} P\{X_2 \in D\} = P\{X_1 + Y \in D\} &= \int P\{X_1 + y \in D\}\,\mathrm{d}P_Y(y) \\ &\leqslant \int P\{X_1 \in D\}\,\mathrm{d}P_Y(y) = P\{X_1 \in D\}. \end{aligned}$$

2.4 Khatri-Šidák Type Inequalities

2.4.a. Let $X^{(1)}$ and $X^{(2)}$ be Gaussian vectors in $\mathbb{R}^m$ and $\mathbb{R}^n$ respectively, and D_1, D_2 be two convex sets in $\mathbb{R}^m$ and $\mathbb{R}^n$ which are symmetric with respect to the origin. If the rank of the covariance matrix $\mathrm{Cov}(X^{(1)}, X^{(2)})$ is not larger than 1, then we have

$$P\{X^{(1)} \in D_1, X^{(2)} \in D_2\} \geqslant P\{X^{(1)} \in D_1\}P\{X^{(2)} \in D_2\}.$$

Proof. Let g and h be two functions defined in $\mathbb{R}^N$. Suppose that for any $x_1, x_2 \in \mathbb{R}^N$, $(g(x_1) - g(x_2))(h(x_1) - h(x_2)) \leqslant 0$. Let X be a random vector in $\mathbb{R}^N$ and Y be an iid. copy of it. Then from $E(g(X) - g(Y))(h(X) - h(Y)) \geqslant 0$ we have

$$Eg(X)h(X) \geqslant Eg(X)Eh(X). \tag{3}$$

Let Σ_1 and Σ_2 be the covariance matrixes of $X^{(1)}$ and $X^{(2)}$ respectively. Since the rank of $\mathrm{Cov}(X^{(1)}, X^{(2)})$ is at most 1, there exist a vector $a \in \mathbb{R}^m$ and a vector $b \in \mathbb{R}^n$ which satisfy $\mathrm{Cov}(X^{(1)}, X^{(2)}) = ab'$, and $X^{(1)}$, $X^{(2)}$ can be expressed as

$$X^{(1)} = Y^{(1)} + aG, \quad X^{(2)} = Y^{(2)} + bG,$$

where G is a standard normal variable, $Y^{(1)}$ and $Y^{(2)}$ are Gaussian vectors with mean 0 and covariance matrices $\Sigma_1 - aa'$ and $\Sigma_2 - bb'$[2] respectively, and $Y^{(1)}$, $Y^{(2)}$ and G are independent of each other. From 2.3.a, we obtain $P\{Y^{(1)} + ay \in D_1\}$ and $P\{Y^{(2)} + by \in D_2\}$ are both nondecreasing functions of $|y|$. Therefore, by using (3) we have

$$\begin{aligned}
&P\{X^{(1)} \in D_1, X^{(2)} \in D_2\} \\
=&P\{Y^{(1)} + aG \in D_1, Y^{(2)} + bG \in D_2\} \\
=&\int P\{Y^{(1)} + ay \in D_1, Y^{(2)} + by \in D_2\}\, \mathrm{d}P_g(y)
\end{aligned}$$

[2] If we arbitrarily choose the vectors a and b for the expression of $\mathrm{Cov}(X^{(1)}, X^{(2)})$, the matrices $\Sigma_1 - aa'$ and $\Sigma_2 - bb'$ may not be non-negatively definite (nnd.) and thus $Y^{(1)}$ and $Y^{(2)}$ are not well defined. We just remind the reader that we can always select suitable vectors a and b such that the matrices $\Sigma_1 - aa'$ and $\Sigma_2 - bb'$ are both nnd. and our proof is based on such a suitable selection of the vectors a and b.

$$
\begin{aligned}
&= \int P\{Y^{(1)} + ay \in D_1\} P\{Y^{(2)} + by \in D_2\} \, \mathrm{d}P_g(y) \\
&\geqslant \int P\{Y^{(1)} + ay \in D_1\} \, \mathrm{d}P_g(y) \int P\{Y^{(2)} + by \in D_2\} \, \mathrm{d}P_g(y) \\
&= P\{Y^{(1)} + ag \in D_1\} P\{Y^{(2)} + bg \in D_2\} \\
&= P\{X^{(1)} \in D_1\} P\{X^{(2)} \in D_2\}.
\end{aligned}
$$

As a special case of 2.4.a, we have

2.4.b. Suppose $(X_1, \cdots, X_N)$ is a zero-mean Gaussian vector in $\mathbb{R}^N$. Then for any positive number $\lambda_i, i = 1, \cdots, N$, we have

$$
\begin{aligned}
P\left\{\bigcap_{i=1}^{N} (|X_i| \leqslant \lambda_i)\right\} &\geqslant P\Big\{\bigcap_{i=1}^{N-1} (|X_i| \leqslant \lambda_i)\Big\} P\{|X_N| \leqslant \lambda_N\} \\
&\geqslant \prod_{i=1}^{N} P\{|X_i| \geqslant \lambda_N\}.
\end{aligned}
$$

When the rank of $\mathrm{Cov}(X^{(1)}, X^{(2)})$ is larger than 1, Shao(2003) proved the following conclusion.

2.4.c. Suppose $(X_1, \cdots, X_n)$ is a zero-mean Gaussian vector. Then for any $x > 0$ and each $1 < k \leqslant n$,

$$
\begin{aligned}
&\rho P\{\max_{1 \leqslant i \leqslant k} |X_i| \leqslant x\} P\{\max_{k < i \leqslant n} |X_i| \leqslant x\} \\
&\leqslant P\{\max_{1 \leqslant i \leqslant n} |X_i| \leqslant x\} \leqslant \rho^{-1} P\{\max_{1 \leqslant i \leqslant k} |X_i| \leqslant x\} P\{\max_{k < i \leqslant n} |X_i| \leqslant x\},
\end{aligned}
$$

where $\rho = (|\Sigma|/(|\Sigma_{11}||\Sigma_{22}|))^{1/2}$, Σ, Σ_{11} and Σ_{22} are covariance matrixes of $(X_1, \cdots, X_n)$, $(X_1, \cdots, X_k)$ and $(X_{k+1}, \cdots, X_n)$ respectively.

2.4.b can be regarded as an analogy of an absolute-value situation of Slepian lemma. Another analogy is 2.5.

2.5 Corner Probability of Normal Vector

Let $X = (X_1, \cdots, X_N)$ be a zero-mean Gaussian vector in $\mathbb{R}^N$ with covariance matrix $\Gamma = (a_{ij})$ satisfying $a_{ij} = \alpha_i \alpha_j (a_{ii} a_{jj})^{1/2}$, $i \neq j$, where $|\alpha_i| \leqslant 1$ and $a_{ii} > 0, i = 1, \cdots, N$. Then for any positive number $\lambda_i, i = 1, \cdots, N$, we have

$$
P\left\{\bigcap_{i=1}^{N} (|X_i| \geqslant \lambda_i)\right\} \geqslant \prod_{i=1}^{N} P\{(|X_i| \geqslant \lambda_i)\}.
$$

Proof. Put $\sigma_i^2 = a_{ii}$. By the assumption, the covariance matrix Γ can be written as $\Gamma = T + \alpha\alpha'$, where $\alpha = (\sigma_1\alpha_1, \cdots, \sigma_N\alpha_N)'$, and T is a $N \times N$ diagonal matrix of which the diagonal elements are $\sigma_i^2(1-\alpha_i^2)$.

Moreover, X can be written as

$$X = Y + \alpha g,$$

where $Y = (Y_1, \cdots, Y_N)$ is a zero-mean Gaussian vector with covariance matrix T, g is a standard normal variable independent of Y, and $Y_1, \cdots, Y_N, g$ are independent of each other. From 2.3.a, we obtain that for each i, $P\{|Y_i + \sigma_i\alpha_i y| \geqslant \lambda_i\}$ is a nondecreasing function of $|y|$. By using (3), we have

$$\begin{aligned} P\left\{\bigcap_{i=1}^N (|X_i| \geqslant \lambda_i)\right\} &= \int \prod_{i=1}^N P\{|Y_i + \sigma_i\alpha_i y| \geqslant \lambda_i\}\, \mathrm{d}P_g(y) \\ &\geqslant \prod_{i=1}^N \int P\{|Y_i + \sigma_i\alpha_i y| \geqslant \lambda_i\}\, \mathrm{d}P_g(y) \\ &= \prod_{i=1}^N P\{|X_i| \geqslant \lambda_i\}. \end{aligned}$$

2.6 Normal Approximations of Binomial and Poisson Distributions

2.6.a (DeMoivre-Laplace). For $n = 1, 2, \cdots$, let $k = k_n$ be a non-negative integer and put $x = x_k = (k-np)(npq)^{-1/2}$, where $q = 1-p, 0 < p < 1$. If $x = o(n^{1/6})$, there exist positive constants A, B, C such that

$$\left|\frac{b(k;n,p)}{(npq)^{-1/2}\varphi(x)} - 1\right| < \frac{A}{n} + \frac{B \mid x\mid^3}{\sqrt{n}} + \frac{C\mid x\mid}{\sqrt{n}}.$$

Proof. The condition $x = o(n^{1/6})$ implies $k/n \to p$. By the Stirling formula

$$n! = n^{n+1/2}\mathrm{e}^{-n+\varepsilon_n}\sqrt{2\pi}, \qquad \frac{1}{12n+1} < \varepsilon_n < \frac{1}{12n},$$

we have

$$\begin{aligned} b(k;n,p) &= \binom{n}{k} p^k q^{n-k} = \frac{n^{n+1/2}\exp(-n+\varepsilon_n)(2\pi)^{-1/2}p^k q^{n-k}}{k^{k+1/2}(n-k)^{n-k+1/2}\exp(-n+\varepsilon_k+\varepsilon_{n-k})} \\ &= \frac{\mathrm{e}^{\varepsilon}}{\sqrt{2\pi}}\left(\frac{k}{np}\right)^{-k-1/2}\left(\frac{n-k}{nq}\right)^{-n+k-1/2}(npq)^{-1/2}, \end{aligned}$$

where $\varepsilon = \varepsilon_n - \varepsilon_k - \varepsilon_{n-k}$. Since $k/n \to p, \varepsilon = O(n^{-1})$. Now

$$\begin{aligned}
&\log\{(2\pi npq)^{1/2} b(k;n,p)\} \\
=&\varepsilon - \left(k+\frac{1}{2}\right)\log\frac{k}{np} - \left(n-k+\frac{1}{2}\right)\log\frac{n-k}{nq} \\
=&\varepsilon - \left(np + x\sqrt{npq} + \frac{1}{2}\right)\log\left(1 + x\sqrt{\frac{q}{np}}\right) \\
&- \left(nq - x\sqrt{npq} + \frac{1}{2}\right)\log\left(1 - x\sqrt{\frac{p}{nq}}\right) \\
=&\varepsilon - \left(np + x\sqrt{npq} + \frac{1}{2}\right)\left[x\sqrt{\frac{q}{np}} - \frac{x^2 q}{2np} + O\left(\frac{|x|^3}{n^{3/2}}\right)\right] \\
&- \left(nq - x\sqrt{npq} + \frac{1}{2}\right)\left[-x\sqrt{\frac{p}{nq}} - \frac{x^2 p}{2nq} + O\left(\frac{|x|^3}{n^{3/2}}\right)\right] \\
=&\varepsilon - \left[x\sqrt{npq} + x^2 q - \frac{x^2 q}{2} + \frac{x}{2}\sqrt{\frac{q}{np}} + O\left(\frac{|x|^3}{n^{1/2}}\right) + O\left(\frac{x^2}{n}\right)\right] \\
&- \left[-x\sqrt{npq} + x^2 p - \frac{x^2 p}{2} - \frac{x}{2}\sqrt{\frac{p}{nq}} + O\left(\frac{|x|^3}{n^{1/2}}\right) + O\left(\frac{x^2}{n}\right)\right] \\
=&-\frac{x^2}{2} + O\left(\frac{|x|^3}{\sqrt{n}}\right) + O\left(\frac{|x|}{\sqrt{n}}\right) + O\left(\frac{1}{n}\right).
\end{aligned}$$

Hence,

$$\begin{aligned}
(npq)^{1/2} b(k;n,p) &= \frac{1}{\sqrt{2\pi}}\exp\left(\frac{-x^2}{2} + O\left(\frac{|x|^3}{\sqrt{n}}\right) + O\left(\frac{|x|}{\sqrt{n}}\right) + O\left(\frac{1}{n}\right)\right) \\
&= \varphi(x)\left[1 + O\left(\frac{|x|^3}{\sqrt{n}}\right) + O\left(\frac{|x|}{\sqrt{n}}\right) + O\left(\frac{1}{n}\right)\right].
\end{aligned}$$

This proves the desired inequality.

2.6.b. For $x = (k-\lambda)/\sqrt{\lambda} \sim o(\lambda^{1/6})$ as $\lambda \to \infty$, there exist positive constants A, B and C such that

$$\left|\frac{p(k;\lambda)}{\lambda^{-1/2}\varphi(x)} - 1\right| \leqslant \frac{A}{\lambda} + \frac{B|x|^3}{\sqrt{\lambda}} + \frac{C|x|}{\sqrt{\lambda}}.$$

Proof. Since $x = (k-\lambda)/\sqrt{\lambda} = o(\lambda^{1/6}), k = \lambda + o(\lambda^{2/3})$ and $x\sqrt{\lambda}/k = o(\lambda^{-1/3})$,

$$\log\left\{(2\pi\lambda)^{1/2}\frac{\lambda^k e^{-\lambda}}{k!}\right\} = \log\left\{\left(\frac{\lambda}{k}\right)^{k+1/2} e^{-\lambda+k-\varepsilon_k}\right\}$$

$$
\begin{aligned}
&= \left(k + \frac{1}{2}\right) \log\left(1 - \frac{x\sqrt{\lambda}}{k}\right) + x\sqrt{\lambda} - \varepsilon_k \\
&= -\left(k + \frac{1}{2}\right)\left\{\frac{x\sqrt{\lambda}}{k} + \frac{x^2\lambda}{2k^2} + O\left(\frac{x^3\lambda^{3/2}}{k^3}\right)\right\} + x\sqrt{\lambda} - \varepsilon_k \\
&= -\left\{\frac{x^2\lambda}{2k} + \frac{x\sqrt{\lambda}}{2k} + \frac{x^2\lambda}{4k^2} + O\left(\frac{x^3\lambda^{3/2}}{k^2}\right)\right\} \\
&= -\frac{1}{2}x^2 + O\left(\frac{x}{\sqrt{\lambda}}\right) + O\left(\frac{1}{\lambda}\right) + O\left(\frac{x^3}{\sqrt{\lambda}}\right),
\end{aligned}
$$

which implies the desired inequality.

References

Anderson T W. 1955. The integral of a symmetric unimodal function over a symmetric convex set and some probability inequalities. Proc. Amer. Soc., 6: 170-176.

Barbour A D, Holst L, Janson S. 1992. Poisson Approximation. Oxford: Clarendon Press,

Berman S M. 1964. Limit theorems for the maximum term in stationary sequences. Ann. Math. Statist., 35: 502-516.

Feller W. 1968. An Introduction to Probability Theory and Its Applications. 3rd ed.. New York: Wiley.

Gordon Y. 1985. Some inequalities for Gaussian processes and applications. Israel J. Math., 50: 265-289.

Khatri G G. 1967. On centain inequalities for normal distributions and the applications to simultaneous confidence bounds. Ann. Math. Statist., 38: 1853-1867.

Kimball A W. 1951. On dependent tests of significance in the analysis of variance. Ann. Math. Statist., 22: 600-602.

Lin Z Y, Lu C L, Zhang L X. 2001. Path Properties of Gaussian Processes. Beijing: Science Press & Hangzhou: Zhejiang Univ. Press.

Loeve M. 1977. Probability Theory. New York: Springer-Verlag.

Shorack G R, Wellner J A. 1986. Empirical processes with applications to statistics. New York: Wiley.

Šidák Z. 1968. On multivariate normal probabilities of rectangales: their dependence on correlation. Ann. Math. Statist., 39: 1425-1434.

Slepian D. 1962. The one-side barrier problem for Gaussian noise. Bell System Tech. J., 41: 463-501.

Tong Y L. 1980. Probability Inequalities in Multivariate Distributions. New York: Academic Press.

Chapter 3

Inequalities Related to Characteristic Functions

The characteristic function (c.f.) or Fourier transformation is an important mathematical tool in probability theory, especially in the theory of limiting theorems of sums of independent random variables. Most of the inequalities can be found in Loéve (1977) and Petrov (1995).

Let ξ be a r.v. with d.f. $F(x)$. Its c.f. is defined as

$$f(x) = E\mathrm{e}^{\mathrm{i}t\xi} = \int_{-\infty}^{\infty} \mathrm{e}^{\mathrm{i}tx}\mathrm{d}F(x).$$

3.1 Inequalities Related Only with c.f.

3.1.a. For any real t,

$$1 - |f(2t)|^2 \leqslant 4(1 - |f(t)|^2).$$

Proof. Let $G(x)$ be an arbitrary d.f. and let $g(t)$ be the corresponding c.f. Then

$$\mathrm{Re}(1 - g(t)) = \int_{-\infty}^{\infty} (1 - \cos tx)\mathrm{d}G(x),$$

where Re denotes the real part. It is clear that

$$1 - \cos tx = 2\sin^2 \frac{tx}{2} \geqslant \frac{1}{4}(1 - \cos 2tx)$$

and therefore for every t,

$$\mathrm{Re}(1 - g(2t)) \leqslant 4\mathrm{Re}(1 - g(t))$$

(this inequality is of its own interest). The desired inequality then follows by setting $g(t) = |f(t)|^2$.

3.1.b. If $|f(t)| \leqslant c < 1$ for all $|t| \in [b, 2b]$ $(b > 0)$, then

$$|f(t)| \leqslant 1 - \frac{1-c^2}{8b^2}t^2$$

for $|t| < b$.

Proof. It follows from 3.1.a that

$$1 - |f(2^n t)|^2 \leqslant 4^n(1 - |f(t)|^2)$$

for every n. For $t = 0$ the inequality is trivial. Suppose $t \neq 0, |t| < b$. We choose n so that $2^{-n}b \leqslant |t| < 2^{-n+1}b$. Then $|f(2^n t)|^2 \leqslant c^2$ and $1 - |f(t)|^2 > \frac{1-c^2}{4b^2}t^2$, or $|f(t)| < 1 - \frac{1-c^2}{8b^2}t^2$.

3.1.c. Let $f(t)$ be the c.f. of a non-degenerate distribution. Then there exist positive constants δ and ε such that $|f(t)| \leqslant 1 - \varepsilon t^2$ for $|t| \leqslant \delta$.

Proof. We first prove that for any non-degenerate distribution with c.f. $f(t)$, there is a constant b such that $|f(t)| < 1$ for any $|t| \leqslant 2b$. If this is proved, then the conclusion follows from 3.1.b by choosing $c = \sup_{t \in [b,2b]} |f(t)|$ which is less than 1 by continuity of the c.f.

Now, we proceed with the proof of our assertion. If there is a positive value t_0 such that $|f(t_0)| = 1$, then there is a real number a such that $f(t_0) = \mathrm{e}^{\mathrm{i}a}$, which implies that $f(t_0)\mathrm{e}^{-\mathrm{i}a} = 1$. Considering its real part we have

$$\int (1 - \cos(t_0 x - a))\mathrm{d}F(x) = 0.$$

Since $1 - \cos(t_0 x - a) \geqslant 0$, we conclude that with probability 1, $t_0 x - a = 2\pi k$, $k = 0, \pm 1, \cdots$. That is, F is a lattice distribution valued at $a + 2\pi k/t_0$. If there is a sequence $t_n \downarrow 0$ such that $|f(t_n)| = 1$, then the r.v. can only take values in

$$\bigcap_{n=1}^{\infty} \{a_n + 2\pi k/t_n, k = 0, \pm 1, \cdots\}.$$

The intersection can contain at most one point by the fact that $t_n \to 0$ and thus F is degenerate which contradicts the assumption. Our assertion then follows from 3.1.b and consequently 3.1.c is proved.

3.1.d. Let ξ be a bounded r.v. with $|\xi| \leqslant M$ and variance σ^2. Then

$$\mathrm{e}^{-\sigma^2 t^2} \leqslant |f(t)| \leqslant \mathrm{e}^{-\sigma^2 t^2/3}, \quad |t| \leqslant \frac{1}{4M}.$$

Proof. Suppose first that $E\xi = 0$. By Taylor's formula

$$f(t) = \sum_{j=0}^{n-1} \frac{(\mathrm{i}t)^j}{j!} E\xi^j + R_n(t), \quad |R_n(t)| \leqslant \frac{|t|^n}{n!} E|\xi|^n.$$

Hence, we have

$$|1 - f(t)| \leqslant \frac{\sigma^2 t^2}{2} \tag{4}$$

and

$$\left|1 - f(t) - \frac{\sigma^2 t^2}{2}\right| \leqslant \frac{M\sigma^2 |t|^3}{6}. \tag{5}$$

If z is a complex number and $|1 - z| < 1$,

$$|\log z + 1 - z| = \left|\int_z^1 \left(\frac{1}{\zeta} - 1\right) \mathrm{d}\zeta\right| \leqslant \frac{|1 - z|^2}{|z|}$$

(integrating along a line segment). Then, using (4), we obtain

$$|\log f(t) + 1 - f(t)| \leqslant \frac{|1 - f(t)|^2}{|f(t)|} \leqslant \frac{\sigma^4 t^4}{4(1 - \sigma^2 t^2/2)}, \quad M^2 t^2 < 2.$$

Combining this inequality with (5), it follows that

$$\left|-\log f(t) - \frac{\sigma^2 t^2}{2}\right| \leqslant \frac{\sigma^4 t^4}{4(1 - \sigma^2 t^2/2)} + \frac{M\sigma^2 |t|^3}{6} \leqslant \frac{\sigma^2 t^2}{6}, \quad M|t| \leqslant \frac{1}{2}.$$

Taking real parts, we find that the desired inequality is true for $|t| \leqslant 1/(2M)$. If we now drop the restriction that $E\xi = 0$, we apply the inequality to $\xi - E\xi$ to obtain 3.1.d.

3.1.e (increment inequality). For all real t and h,

$$|f(t) - f(t + h)|^2 \leqslant 2(1 - \mathrm{Re} f(h)).$$

Proof. By the Cauchy-Schwarz inequality

$$\begin{aligned}
|f(t) - f(t + h)|^2 &= \left|\int \mathrm{e}^{\mathrm{i}tx}(1 - \mathrm{e}^{\mathrm{i}hx}) \mathrm{d}F(x)\right|^2 \\
&\leqslant \int \mathrm{d}F(x) \int |1 - \mathrm{e}^{\mathrm{i}hx}|^2 \mathrm{d}F(x) \\
&= 2\int (1 - \cos hx) \mathrm{d}F(x) \\
&= 2(1 - \mathrm{Re} f(h)).
\end{aligned}$$

3.2 Inequalities Related to c.f. and d.f.

3.2.a (truncation inequality). For $u > 0$,

$$\int_{|x|<1/u} x^2 \mathrm{d}F(x) \leqslant \frac{3}{u^2}(1 - \mathrm{Re} f(u)),$$

$$\int_{|x|\geqslant 1/u} \mathrm{d}F(x) \leqslant \frac{7}{u}\int_0^u (1 - \mathrm{Re} f(t))\mathrm{d}t.$$

Proof.

$$\begin{aligned}\int (1-\cos ux)\mathrm{d}F(x) &\geqslant \int_{|x|<1/u} \frac{u^2x^2}{2}\left(1-\frac{u^2x^2}{12}\right)\mathrm{d}F(x)\\ &\geqslant \frac{11u^2}{24}\int_{|x|<1/u} x^2\mathrm{d}F(x);\end{aligned}$$

$$\begin{aligned}\frac{1}{u}\int_0^u \mathrm{d}t \int (1-\cos tx)\mathrm{d}F(x) &= \int \left(1-\frac{\sin ux}{ux}\right)\mathrm{d}F(x)\\ &\geqslant (1-\sin 1)\int_{|x|\geqslant 1/u} \mathrm{d}F(x).\end{aligned}$$

3.2.b (integral inequality). For $u > 0$ there exist functions $0 < m(u) < M(u) < \infty$ such that

$$m(u)\int_0^u (1-\mathrm{Re} f(t))\mathrm{d}t \leqslant \int \frac{x^2}{1+x^2}\mathrm{d}F(x) \leqslant M(u)\int_0^u (1-\mathrm{Re} f(t))\mathrm{d}t.$$

For u sufficiently close to 0,

$$\int \frac{x^2}{1+x^2}\mathrm{d}F(x) \leqslant -M(u)\int_0^u (\log \mathrm{Re} f(t))\mathrm{d}t.$$

Proof. The first part follows from the facts

$$\int_0^u \mathrm{d}t\int (1-\cos tx)\mathrm{d}F(x) = u\int\left(1-\frac{\sin ux}{ux}\right)\frac{1+x^2}{x^2}\frac{x^2}{1+x^2}\mathrm{d}F(x)$$

and

$$0 < m^{-1}(u) \leqslant |u|\left(1-\frac{\sin ux}{ux}\right)\frac{1+x^2}{x^2} \leqslant M^{-1}(u) < \infty.$$

The second part follows from the fact that $\ln(1-x) \sim x$ as $x \to 0$.

3.2.c. $\int \frac{x^2}{1+x^2}\mathrm{d}F(x) \leqslant \int_0^\infty \mathrm{e}^{-t}|1-f(t)|\mathrm{d}t.$

Proof. Integrating by parts yields

$$\int_0^\infty \mathrm{e}^{-t}\cos xt\mathrm{d}t = \frac{1}{1+x^2},$$

which implies

$$\int \frac{x^2}{1+x^2}\mathrm{d}F(x) = \int_0^\infty \mathrm{e}^{-t}(1-\mathrm{Re}f(t))\mathrm{d}t,$$

as required.

3.3 Normality Approximations of c.f. of Independent Sums

Let $X_1, \cdots, X_n$ be independent r.v.'s, $EX_j = 0$, $E|X_j|^3 < \infty$, $j = 1, \cdots, n$. Put

$$\sigma_j^2 = EX_j^2, \quad B_n = \sum_{j=1}^n \sigma_j^2, \quad L_n = B_n^{-3/2}\sum_{j=1}^n E|X_j|^3.$$

Let $f_n(t)$ be the c.f. of the r.v. $B_n^{-1/2}\sum\limits_{j=1}^n X_j$. Then

$$|f_n(t) - \mathrm{e}^{-t^2/2}| \leqslant 16L_n|t|^3\mathrm{e}^{-t^2/3}$$

for $|t| \leqslant \frac{1}{4L_n}$.

Proof. We begin with the case in which $|t| \geqslant \frac{1}{2}L_n^{-1/3}$. Then $8L_n|t|^3 \geqslant 1$, and we will show that

$$|f_n(t)|^2 \leqslant \mathrm{e}^{-2t^2/3}, \tag{6}$$

which implies that

$$|f_n(t) - \mathrm{e}^{-t^2/2}| \leqslant |f_n(t)| + \mathrm{e}^{-t^2/2} \leqslant 2\mathrm{e}^{-t^2/3} \leqslant 16L_n|t|^3\mathrm{e}^{-t^2/3}.$$

Write $v_j(t) = E\mathrm{e}^{\mathrm{i}tX_j}$ $(j = 1, \cdots, n)$ and define the symmetrization r.v. $\tilde{X}_j = X_j - Y_j$, where Y_j is independent and identically distributed as X_j. Then, $\tilde{X}_j$ has the c.f. $|v_j(t)|^2$ and variance $2\sigma_j^2$. Furthermore, $E|\tilde{X}_j|^3 \leqslant 8E|X_j|^3$,

$$|v_j(t)|^2 \leqslant 1 - \sigma_j^2t^2 + \frac{4}{3}|t|^3E|X_j|^3 \leqslant \exp\left\{-\sigma_j^2t^2 + \frac{4}{3}|t|^3E|X_j|^3\right\}.$$

Therefore in the interval $|t| \leqslant \frac{1}{4L_n}$ we have the estimate

$$|f_n(t)|^2 = \prod_{j=1}^{n} \left| v_j \left(\frac{t}{\sqrt{B_n}} \right) \right|^2 \leqslant \exp\left\{ -t^2 + \frac{4}{3} L_n |t|^3 \right\} \leqslant \exp\left\{ -\frac{2}{3} t^2 \right\},$$

and (6) is proved.

Now suppose that $|t| \leqslant \frac{1}{4L_n}$ and $|t| < \frac{1}{2} L_n^{-1/3}$. For $j = 1, \cdots, n$, we have

$$\frac{\sigma_j}{\sqrt{B_n}} |t| \leqslant \frac{(E|X_j|^3)^{1/3}}{\sqrt{B_n}} |t| < L_n^{1/3} |t| < \frac{1}{2}, \quad v_j \left(\frac{t}{\sqrt{B_n}} \right) = 1 - r_j,$$

where

$$r_j = \frac{\sigma_j^2 t^2}{2B_n} + \theta_j \frac{E|X_j|^3}{6B_n^{3/2}} |t|^3, \quad |\theta_j| \leqslant 1,$$

so that $|r_j| < \frac{1}{6}$ and

$$|r_j|^2 \leqslant 2 \left(\frac{\sigma_j^2 t^2}{2B_n} \right)^2 + 2 \left(\frac{E|X_j|^3}{6B_n^{3/2}} |t|^3 \right)^2 \leqslant \frac{E|X_j|^3}{3B_n^{3/2}} |t|^3.$$

Therefore

$$\log v_j \left(\frac{t}{\sqrt{B_n}} \right) = -\frac{\sigma_j^2 t^2}{2B_n} + \theta'_j \frac{E|X_j|^3}{2B_n^{3/2}} |t|^3, \quad |\theta'_j| \leqslant 1,$$

$$\log f_n(t) = -\frac{t^2}{2} + \theta \frac{L_n}{2} |t|^3, \quad |\theta| \leqslant 1.$$

Using the inequality $L_n |t|^3 < \frac{1}{8}$, which implies that $\exp\left\{ \frac{1}{2} L_n |t|^3 \right\} < 2$, we find that

$$\begin{aligned} |f_n(t) - \mathrm{e}^{-t^2/2}| &\leqslant \mathrm{e}^{-t^2/2} \left| \mathrm{e}^{\frac{\theta}{2} L_n |t|^3} - 1 \right| \leqslant \frac{L_n}{2} |t|^3 \exp\left\{ -\frac{t^2}{2} + \frac{L_n}{2} |t|^3 \right\} \\ &\leqslant L_n |t|^3 \mathrm{e}^{-t^2/2}. \end{aligned}$$

References

Loéve M. 1977. Probability Theory. 4th ed.. New York: Springer-Verlag.

Petrov V V. 1995. Limit Theorems of Probability Theory: Sequences of Independent Random Variables. Oxford & New York: Clarendon Press & Oxford University Press.

Chapter 4

Estimates of the Difference of Two Distribution Functions

Rates of weak convergence are important for application of weak limit theorems. Thus, investigation on convergence rates has been an active research topic for decades. Generally speaking, the convergence rates are established by various basic inequalities between two distribution functions and/or functions of bounded variation in terms of various transformations. The first work was done be Berry-Esseen who established the convergence rate of normal approximation in terms of Fourier transformations or characteristic functions. Stein and Chen created a new method to evaluate the convergence rates of normal or Poisson approximation for non-independent sums. In 1993, Bai established convergence rates of empirical spectral distributions of large dimensional random matrices in terms of Stieltjes transforms. In this chapter, we only introduce some basic inequalities of difference of two distribution functions. Their applications can be found in Petrov (1995), Stein (1986) and Bai (1993).

4.1 Fourier Transformation

4.1.a (Berry-Esseen basic inequality). Estimate of the difference of the corresponding c.f.'s.

Let $F(x)$ be a non-decreasing bounded function, and $G(x)$ a function of bounded variation on the real line. Suppose that $F(-\infty) = G(-\infty)$. Let

$$f(t) = \int \mathrm{e}^{\mathrm{i}tx} \mathrm{d}F(x), \quad g(t) = \int \mathrm{e}^{\mathrm{i}tx} \mathrm{d}G(x),$$

and T be an arbitrary positive number. Then for any $b > \frac{1}{2\pi}$ we have

$$\sup_{-\infty<x<\infty} |F(x)-G(x)| \leqslant b\int_{-T}^{T}\left|\frac{f(t)-g(t)}{t}\right| dt + 2bT \sup_{-\infty<x<\infty}\int_{|y|\leqslant c(b)/T} |G(x+y)-G(x)|dy,$$

where $c(b)$ is a positive constant depending only on b and is usually chosen as the root of the equation

$$\int_0^{c(b)/2} \frac{\sin^2 x}{x^2}dx = \frac{\pi}{4}+\frac{1}{8b}.$$

Proof. Note that $w(x) = \frac{\sin^2 x}{\pi x^2}$ is a probability density function with c.f. $h(t) = (1-|t|)$ or 0 according to whether $|t|<1$ or not. Let $\tilde{F}$ ($\tilde{G}$) be the convolution of $Tw(Tx)$ with F (G, correspondingly). Then, we have

$$\tilde{F}(x)-\tilde{G}(x) = \frac{1}{2\pi}\int_{-\infty}^{\infty} e^{-itx}\frac{f(t)-g(t)}{-it}h(t/T)dt,$$

from which it follows that

$$\sup_x |\tilde{F}(x)-\tilde{G}(x)| \leqslant \frac{1}{2\pi}\int_{-T}^{T}\frac{|f(t)-g(t)|}{|t|}dt. \tag{7}$$

On the other hand, we have

$$\tilde{F}(x)-\tilde{G}(x) = \int_{-\infty}^{\infty} w(y)[F(x-y/T)-G(x-y/T)]dy.$$

Let $\Delta = \sup_x |F(x)-G(x)|$. There exists x_0 such that either $F(x_0) - G(x_0 \pm 0) = \Delta$ or $G(x_0 \pm 0) - F(x_0 - 0) = \Delta$. We consider the first case. Then, we have

$$\begin{aligned}
&\tilde{F}\left(x_0+\frac{c(b)}{2T}\right)-\tilde{G}\left(x_0+\frac{c(b)}{2T}\right)\\
&=\int_{-\infty}^{\infty} w(y)\left[F\left(x_0-\frac{y-c(b)/2}{T}\right)-G\left(x_0-\frac{y-c(b)/2}{T}\right)\right]dy\\
&\geqslant \int_{|y|<c(b)/2} w(y)\left[F\left(x_0-\frac{y-c(b)/2}{T}\right)-G\left(x_0-\frac{y-c(b)/2}{T}\right)\right]dy\\
&\quad -\Delta\int_{|y|>c(b)/2} w(y)dy
\end{aligned}$$

$$\geqslant -\int_{|y|<c(b)/2} \pi^{-1} \left| G(x_0 \pm 0) - G\left(x_0 - \frac{y - c(b)/2}{T}\right) \right| \mathrm{d}y$$
$$+\Delta \left(1 - 2\int_{|y|>c(b)/2} w(y)\mathrm{d}y\right)$$
$$\geqslant \frac{\Delta}{2\pi b} - \pi^{-1} T \sup_x \int_{|y|<c(b)/T} |G(x) - G(x-y)|\mathrm{d}y.$$

The Berry Esseen inequality follows by substituting the above into (7).

The proof for the second case is similar and therefore is omitted.

Remark. By the difference of the corresponding c.f.'s, this inequality is to be used to establish the convergence rate for normal approximation or Edgeworth expansions for sums of independent r.v.'s. Most of the key inequalities are due to Berry and Esseen and can be found in Loéve (1977) or Petrov (1995).

4.1.b (Esseen and Berry-Esseen inequalities). Let $X_1, \cdots, X_n$ be independent r.v.'s with $EX_j = 0, E|X_j| < \infty,\ j = 1, \cdots, n$. Put

$$\sigma_j^2 = EX_j^2, \quad B_n = \sum_{j=1}^n \sigma_j^2, \quad F_n(x) = P\left(B_n^{-1/2} \sum_{j=1}^n x_j < x\right),$$

$$L_n = B_n^{-3/2} \sum_{j=1}^n E|X_j|^3.$$

Then there exists a constant $A_1 > 0$ such that

$$\Delta_n \equiv \sup_{-\infty<x<\infty} |F_n(x) - \Phi(x)| \leqslant A_1 L_n$$

(Esseen's inequality). Specially, if $X_1, \cdots, X_n$ are independent identically distributed (i.i.d.) with $\sigma^2 = EX_1^2, \rho = E|X_1|^3/\sigma^3$, there exists a constant $A_2 > 0$ such that

$$\Delta_n \leqslant A_2 \rho / \sqrt{n}$$

(Berry - Esseen's inequality).

Remark. Here $A_1 \leqslant 0.7915, A_2 \leqslant 0.7655$.

Proof. The d.f.'s $F_n(x)$ and $\Phi(x)$ satisfy the conditions of 4.1.a and $\sup_x |\Phi'(x)| \leqslant 1/\sqrt{2\pi}$. Putting $b = 1/\pi, T = 1/(4L_n)$, we find that

$$\Delta_n \leqslant \frac{1}{\pi} \int_{|t|\leqslant 1/(4L_n)} \left| \frac{f_n(t) - \mathrm{e}^{-t^2/2}}{t} \right| \mathrm{d}t + \frac{1}{4\pi L_n} \int_{|y|\leqslant 4L_n c(1/\pi)} \frac{1}{\sqrt{2\pi}} |y| \mathrm{d}y,$$

which, together with 3.3, implies Esseen's inequality.

4.1.c (generalizations of Berry-Esseen's inequality). Let $X_1, \cdots, X_n$ be i.i.d.r.v.'s with $EX_1 = 0, EX_1^2 = \sigma^2$. Then there exist constants $c_1, c_2 > 0$ such that

$$\left\| P\left\{ \frac{1}{\sigma\sqrt{n}} \sum_{j=1}^{n} X_j < x \right\} - \Phi(x) \right\|_p \leqslant c_1(\delta(n) + n^{-1/2}),$$

$$\left\| P\left\{ \frac{1}{\sigma\sqrt{n}} \sum_{j=1}^{n} X_j < x \right\} - \Phi(x) \right\|_p + n^{-1/2} \geqslant c_2\delta(n),$$

where

$$\delta(n) = EX_1^2 I(|X_1| \geqslant \sqrt{n}) + n^{-1/2} E|X_1|^3 I(|X_1| \leqslant \sqrt{n}) + n^{-1} EX_1^4 I(|X_1| \leqslant \sqrt{n}),$$

$$\|f(x)\|_p = \begin{cases} \sup\limits_{-\infty<x<\infty} |f(x)|, & \text{if } \ p = \infty, \\ \left(\int_{-\infty}^{\infty} |f(x)|^p \mathrm{d}x \right)^{1/p}, & \text{if } \ 1 \leqslant p < \infty. \end{cases}$$

For the proofs, refer to Hall (1982).

Remark. The inequalities can be extended to the non-identically distributed case.

4.1.d (non-uniform estimates). Let $X_1, \cdots, X_n$ be independent r.v.'s satisfying the conditions in 4.2.a. Then there exists a constant $C > 0$ such that

$$|F_n(x) - \Phi(x)| \leqslant C \sum_{i=1}^{n} E|X_i|^3 / (B_n^3(1 + |x|^3)).$$

For the proof, refer to Bikelis (1966).

4.1.e (approximation of moments, von Bahr). Let $X_1, \cdots, X_n$ be independent r.v.'s with $EX_j = 0, E|X_j|^r < \infty$ for some $r > 2, j = 1, \cdots, n$. Put $s_n^2 = \sum\limits_{j=1}^{n} EX_j^2$. Then there exists a constant $M > 0$, such that

$$|E|S_n/s_n|^r - E|N(0,1)|^r| \leqslant Mn^{-(1\wedge(r-2))/2}.$$

If r is an integer $\geqslant 4$ (or $\geqslant 3$ for the i.i.d. case), then the absolute moments can be replaced by moments.

For the proof, refer to von Bahr (1965).

4.2 Stein-Chen Method

This is an alternative method avoiding the use of c.f.'s for establishing the convergence rate for the Poisson approximation for sums of independent or weakly dependent integer-valued r.v.'s.

Let us consider the following example. Suppose $e_1, \cdots, e_n$ are independent r.v.'s with $P(e_j = 1) = p_j, P(e_j = 0) = q_j = 1-p_j, j = 1, \cdots, n$. Put $S_n = \sum_{j=1}^{n} e_j$. Denote the distribution of S_n by $\mathscr{L}_{S_n}$ and let P_λ denote a Poisson distribution with the parameter $\lambda = \sum_{i=1}^{n} p_i$. Let $\mathbb{Z}^+$ be the collection of non-negative integers. Then the total variation distance between $\mathscr{L}_{S_n}$ and P_λ satisfies

$$d_{TV}(\mathscr{L}_{S_n}, P_\lambda) \equiv \sup\{|\mathscr{L}_{S_n}(A) - P_\lambda(A)| : A \subset \mathbb{Z}^+\} \leqslant (1 \wedge \lambda^{-1}) \sum_{j=1}^{n} p_j^2.$$

Proof. For any $A \subset \mathbb{Z}^+$, let function $g = g_{\lambda,A} : \mathbb{Z}^+ \to \mathbb{R}$ be the solution to the difference equation:

$$\lambda g(j+1) - jg(j) = I_A(j) - P_\lambda(A), \quad j \geqslant 0. \tag{8}$$

The value $g(0)$ is not unique but irrelevant, and is conventionally taken as zero. The solution of (8) is easily accomplished recursively, starting with $j = 0$. Substituting $j = S_n$ and taking expectation, we obtain

$$P(S_n \in A) - P_\lambda(A) = E\{\lambda g(S_n + 1) - S_n g(S_n)\}, \tag{9}$$

from which the total variation distance between $\mathscr{L}_{S_n}$ and P_λ can be found, provided that the RHS of (9) can be uniformly estimated for all the $g_{\lambda,A}$. To see how this works, write

$$E(e_j g(S_n)) = E(e_j g(S_n^j + 1)) = p_j E g(S_n^j + 1),$$

where $S_n^j = \sum_{i \neq j} e_i$, because of the independence of e_j and S_n^j. Thus

$$E\{\lambda g(S_n + 1) - S_n g(S_n)\} = \sum_{j=1}^{n} p_j E\{g(S_n + 1) - g(S_n^j + 1)\},$$

and since S_n and S_n^j are equal unless $e_j = 1$, an event of probability p_j, it follows that

$$|P(S_n \in A) - P_\lambda(A)| \leqslant \sup_{j \geqslant 1} |g_{\lambda,A}(j+1) - g_{\lambda,A}(j)| \sum_{j=1}^{n} p_j^2. \tag{10}$$

We now estimate the supremum in the above inequality. Put $U_m = \{0, 1, \cdots, m\}$. Then it is easy to verify that the solution $g = g_{\lambda,A}$ to (8) is given by

$$\begin{aligned} g(j+1) &= \lambda^{-j-1} j! e^{\lambda} \{P_\lambda \left(A \bigcap U_j\right) - P_\lambda(A) P_\lambda(U_j)\} \\ &= \lambda^{-j-1} j! e^{\lambda} \{P_\lambda \left(A \bigcap U_j\right) P_\lambda(U_j^c) \qquad (11) \\ &\quad - P_\lambda \left(A \bigcap U_j^c\right) P_\lambda(U_j)\}, \quad j \geqslant 0. \qquad (12) \end{aligned}$$

Note that the solution to (8) shows that $g_{\lambda,A} = \sum_{i \in A} g_{\lambda,\{i\}}$. Taking $A = \{i\}$, it follows from (12) that $g(j+1)$ is negative and decreasing in j for $j < i$, and positive and decreasing in j for $j \geqslant i$, so that the only positive value of $g_{\lambda,\{i\}}(j+1) - g_{\lambda,\{i\}}(j)$ reaches at $j = i$. Therefore,

$$\begin{aligned} g_{\lambda,A}(j+1) - g_{\lambda,A}(j) &\leqslant g_{\lambda,\{j\}}(j+1) - g_{\lambda,\{j\}}(j) \\ &= e^{-\lambda} \lambda^{-1} \left\{ \sum_{r=j+1}^{\infty} (\lambda^r / r!) + \sum_{r=1}^{j} (\lambda^r / r!) \frac{r}{j} \right\} \\ &\leqslant \lambda^{-1}(1 - e^{-\lambda}). \end{aligned}$$

On the other hand, we have

$$\begin{aligned} g_{\lambda,A}(j+1) - g_{\lambda,A}(j) &\geqslant \sum_{i \geqslant 0, \neq j} g_{\lambda,\{i\}}(j+1) - g_{\lambda,\{i\}}(j) \\ &= -[g_{\lambda,\{j\}}(j+1) - g_{\lambda,\{j\}}(j)] \\ &\geqslant -\lambda^{-1}(1 - e^{-\lambda}). \end{aligned}$$

Combining the two inequalities, it follows that

$$\sup_{j \geqslant 1} |g_{\lambda,A}(j+1) - g_{\lambda,A}(j)| \leqslant \lambda^{-1}(1 - e^{-\lambda}) \leqslant 1 \wedge \lambda^{-1}.$$

Inserting it into (10), the desired estimate follows.

4.3 Stieltjes Transformation

In this section, we establish bounds for difference of d.f.'s in terms of their corresponding Stieltjes transforms. For a bounded variation function $G(x)$ on the real line, define its Stieltjes transform by

$$m_G(z) = \int \frac{1}{x - z} dF(x),$$

where $z = u + \mathrm{i}v$ is a complex variable and $v > 0$. Let $F(x)$ be a d.f. satisfying
$\int |F(x) - G(x)|\mathrm{d}x < \infty$. Then

$$\sup_{-\infty<x<\infty} |F(x) - G(x)| \leqslant \frac{1}{(2\gamma - 1)\pi} \left\{ \int |m_F(z) - m_G(z)|\mathrm{d}u + \frac{1}{v} \sup_{-\infty<x<\infty} \int_{|y|\leqslant 2va} |G(x + y) - G(x)|\mathrm{d}y \right\},$$

where $\gamma > 1/2$ is defined by

$$\gamma = \frac{1}{\pi} \int_{|x|<a} \frac{1}{1 + x^2}\mathrm{d}x.$$

Proof. We have

$$\begin{aligned}
&\pi^{-1} \int |m_F(z) - m_G(z)|\mathrm{d}u \geqslant \pi^{-1} \int_{-\infty}^{x} \mathrm{Im}(m_F(z) - m_G(z))\mathrm{d}u \\
=&\pi^{-1} \int_{-\infty}^{x} \int_{-\infty}^{\infty} \frac{v}{(u - y)^2 + v^2}\mathrm{d}(F(y) - G(y))\mathrm{d}u \\
=&\pi^{-1} \int_{-\infty}^{\infty} \int_{-\infty}^{(x-y)/v} \frac{1}{u^2 + 1}\mathrm{d}(F(y) - G(y))\mathrm{d}u \\
=&\pi^{-1} \int_{-\infty}^{\infty} \int_{-\infty}^{x-uv} \frac{1}{u^2 + 1}\mathrm{d}(F(y) - G(y))\mathrm{d}u \\
=&\pi^{-1} \int_{-\infty}^{\infty} \frac{F(x - uv) - G(x - uv)}{u^2 + 1}\mathrm{d}u \\
\geqslant&\pi^{-1} \int_{|u|<a} \frac{F(x - uv) - G(x - uv)}{u^2 + 1}\mathrm{d}u - \Delta(1 - \gamma) \\
\geqslant&\gamma[F(x - av) - G(x - av)] \\
&-\pi^{-1} \int_{|u|<a} \frac{G(x - av) - G(x - uv)}{u^2 + 1}\mathrm{d}u - \Delta(1 - \gamma) \\
\geqslant&\gamma[F(x - av) - G(x - av)] \\
&-\pi^{-1} \sup_x \int_{|u|<2a} |G(x) - G(x - u)|\mathrm{d}u - \Delta(1 - \gamma).
\end{aligned}$$

Thus we obtain

$$\begin{aligned}
&\sup_x \gamma[F(x) - G(x)] - \Delta(1 - \gamma) \\
\leqslant&\pi^{-1} \int |m_F(z) - m_G(z)|\mathrm{d}u - \pi^{-1} \sup_x \int_{|u|<2a} |G(x) - G(x - u)|\mathrm{d}u. \quad (13)
\end{aligned}$$

Similarly, we have

$$
\begin{aligned}
&\pi^{-1}\int |m_F(z)-m_G(z)|\mathrm{d}u \geqslant \pi^{-1}\int_{-\infty}^{x} \mathrm{Im}(m_G(z)-m_F(z))\mathrm{d}u\\
&\geqslant \gamma[G(x+av)-F(x+av)]\\
&\quad -\pi^{-1}\sup_x \int_{|u|<2a} |G(x)-G(x-u)|\mathrm{d}u-\Delta(1-\gamma).
\end{aligned}
$$

Hence,

$$
\begin{aligned}
&\sup_x \gamma[G(x)-F(x)]-\Delta(1-\gamma)\\
&\leqslant \pi^{-1}\int |m_F(z)-m_G(z)|\mathrm{d}u-\pi^{-1}\sup_x \int_{|u|<2a} |G(x)-G(x-u)|\mathrm{d}u. \quad (14)
\end{aligned}
$$

Combining (13) and (14), we obtain

$$
\Delta(2\gamma-1)\leqslant \pi^{-1}\int |m_F(z)-m_G(z)|\mathrm{d}u-\pi^{-1}\sup_x \int_{|u|<2a} |G(x)-G(x-u)|\mathrm{d}u.
$$

The desired inequality follows.

Remark. By the difference of the corresponding Stieltjes transforms, this inequality is to be used to establish the convergence rate for empirical spectral distributions of large dimensional random matrices.

References

Bai Z D. 1993. Convergence rate of expected spectral distributions of large random matrices. part I. Wigner Matrices. Ann. Probab., 21: 625-648.

Bikelis A. 1966. Estimates of the remainder term in the central limit theorem. Litovsk. Mat. Sb., 6: 323-346.

Hall P. 1982. Rates of Convergence in the Central Limit Theorem. Boston: Pitman.

Loéve M. 1977. Probability Theory. 4th ed.. New York: Springer-Verlag.

Petrov V V. 1995. Limit Theorems of Probability Theory: Sequences of Independent Random Variables. Oxford & New York: Clarendon Press & Oxford University Press.

Shorack G R, Wellner J A. 1986. Empirical processes with applications to statistics. New York: Wiley.

Stein C. 1986. Approximate Computation of Expectations. IMS Lecture Notes-Moograph, Ser. 7.

von Bahr B. 1965. On the convergence in the central limit theorem. Ann. Math. Statist., 36: 808-818.

Chapter 5

Probability Inequalities of Random Variables

Probability inequalities of random variables, especially those of sums of random variables, play important roles in analytic probability theory, say, limiting theorems. Not only the applications of these inequalities, the proofs of these inequalities are also good illustrations of some important mathematical methodologies in probability theory.

Most of the inequalities and their proofs can be found in Loéve (1977), Hall and Heyde (1980). Some new inequalities will be referenced therein.

5.1 Inequalities Related to Two r.v.'s

5.1.a. For any two r.v.'s X and Y,

$$\begin{aligned} &P(X+Y \geqslant x) \leqslant P(X \geqslant x/2) + P(Y \geqslant x/2); \\ &P(|X+Y| \geqslant x) \leqslant P(|X| \geqslant x/2) + P(|Y| \geqslant x/2); \\ &|P(X < x_1, Y < y_1) - P(X < x_2, Y < y_2)| \\ \leqslant &|P(X < x_1) - P(X < x_2)| + |P(Y < y_1) - P(Y < y_2)|. \end{aligned}$$

5.1.b. Suppose that X and Y are independent. Then

$$\begin{aligned} P(X+Y \leqslant x) &\geqslant P(X \leqslant x/2)P(Y \leqslant x/2); \\ P(|X+Y| \leqslant x) &\geqslant P(|X| \leqslant x/2)P(|Y| \leqslant x/2) \end{aligned}$$

and for $x > 0$ large enough,

$$P(|X| > x) \leqslant 2P(|X| > x, |Y| < x/2) \leqslant 2P(|X+Y| > x/2).$$

5.1.c. Let X and Y be i.i.d. r.v.'s. Then for any $a > 0$.

$$P(|X+Y| \leqslant a) \leqslant 2P(|X-Y| \leqslant a).$$

Proof. At first, without loss of generality, we may prove the inequality only for $a = 1$. Furthermore, the inequality follows from

$$P(|X+Y| \leqslant 1) \leqslant \gamma P(|X-Y| \leqslant 1), \quad \forall \gamma > 2. \tag{14}$$

Now, let $\gamma > 2$ be a given constant. We first show that for any probability measure P following which X and Y are distributed,

$$P(\{x; P(|x+Y| \leqslant 1) - \gamma P(|x-Y| \leqslant 1) < 0\}) > 0. \tag{15}$$

If (15) is not true, then $P(A) = 1$, where $A = \{x; P(|x+Y| \leqslant 1) \geqslant \gamma P(|x-Y| \leqslant 1)\}$. Define $\alpha = \sup_{x\in A} P(|x-Y| \leqslant 1)$. Choose as positive $\varepsilon < \gamma^{-2}(\gamma-2)(\gamma-1)\alpha$. Then there is a point $x_0 \in A$ such that $P(|x_0-Y| \leqslant 1) \geqslant \alpha-\varepsilon$. Consequently, we have $P(|x_0+Y| \leqslant 1) \geqslant \gamma\alpha - \gamma\varepsilon$. If there is a point $x_1 \in A\bigcap[-x_0, -x_0+1]$, then $P([-x_0, -x_0+1]) \leqslant P([x_1-1, x_1+1]) \leqslant \alpha$. Therefore, $P([-x_0-1, -x_0]) \geqslant (\gamma-1)\alpha - \gamma\varepsilon$. If such an x_1 does not exist, then $P([-x_0, -x_0+1]) = 0$ and we also have $P([-x_0-1, -x_0]) \geqslant (\gamma-1)\alpha - \gamma\varepsilon$. Finally, we can choose a point $x_2 \in A\bigcap[-x_0-1, -x_0]$ such that $P([x_2-1, x_2+1]) \geqslant P([-x_0-1, -x_0]) \geqslant (\gamma-1)\alpha - \gamma\varepsilon$.

Furthermore, $P([-x_2-1, -x_2+1]) \geqslant \gamma(\gamma-1)\alpha - \gamma^2\varepsilon$. Using the above argument, we may choose a point $x_3 \in A\bigcap[-x_2-1, -x_2]$ such that $P([x_3-1, x_3+1]) \geqslant (\gamma(\gamma-1)-1)\alpha - \gamma^2\varepsilon > \alpha$. This leads to a contradiction to the definition of α. The proof of (15) is now complete.

Finally, let us prove (14). We first consider the discrete case. Suppose that X takes value a_i with probability q_i, $i = 1, \cdots, n$ (note that $\sum q_i = 1$). In this case, we have

$$\gamma P(|X-Y| \leqslant 1) - P(|X+Y| \leqslant 1) = q'Qq,$$

where $q = (q_1, \cdots, q_n)$, $Q = (\gamma I(|a_i - a_j| \leqslant 1) - I(|a_i + a_j| \leqslant 1))_{i,j=1}^n$. By the method of Lagrange multiplier, the minimum of $q'Qq$ is reached when $2Qq - tl = 0$, where l is the vector of all entries 1. By (15), Qq has at least one positive element. Thus all elements are positive and hence $q'Qq > 0$. This proves (14). The general case follows by a routine approach.

Remark. The constant 2 is sharp by the following example. Let X, Y take values $\{-2n+1, \cdots, -1, 2, 4, \cdots, 2n\}$ with probability $1/(2n)$.

Then $P(|X-Y|\leqslant 1)=P(X=Y)=1/(2n)$. On the other hand, $P(|X+Y|\leqslant 1)=(4n-2)/(2n)^2=1/n-1/(2n^2)$.

Although the constant 2 in 5.1.c cannot be further reduced, we have the following integrated inequality.

5.1.d. Let X and Y be i.i.d. r.v.'s. Then for any $a>0$,

$$\int_0^a P(|X+Y|\leqslant x)\mathrm{d}x\leqslant\int_0^a P(|X-Y|\leqslant x)\mathrm{d}x.$$

The conclusion 5.1.d follows easily from the formula

$$\int_0^a P(|X|\leqslant x)\mathrm{d}x=\lim_{T\to\infty}\int_{-T}^{T}\frac{1-\cos(at)}{\pi t^2}f(t)\mathrm{d}t,$$

where f is the c.f. of X.

Remark. As a consequence of 5.1.d, we have the following expectation inequality. Let X and Y be i.i.d. r.v.'s. Then $E|X-Y|\leqslant E|X+Y|$.

Proof.

$$\begin{aligned}&E|X+Y|-E|X-Y|\\&=\lim_{T\to\infty}\int_0^T(P(|X+Y|>x)-P(|X-Y|>x))\mathrm{d}x\\&=\lim_{T\to\infty}\int_0^T(P(|X-Y|\leqslant x)-P(|X+Y|\leqslant x))\mathrm{d}x\geqslant 0.\end{aligned}$$

An alternative proof is to use the formula

$$E|X+Y|-E|X-Y|=\int_0^\infty(1-F(u)-F(-u))(1-G(u)-G(-u))\mathrm{d}u,$$

where F and G are the distribution functions of X and Y respectively.

5.2 Perturbation Inequality

Let $\{X_n,n\geqslant 1\}$ and $\{Y_n,n\geqslant 1\}$ be sequences of r.v.'s such that either (i) X_n and $(Y_1,\cdots,Y_n)$ are independent for all $n\geqslant 1$ or (ii) X_n and $(Y_n,Y_{n+1},\cdots)$ are independent for all $n\geqslant 1$. Then for any constants ε_n,δ_n, ε and δ,

$$P\left\{\bigcup_{n=1}^{\infty}(X_n+Y_n>\varepsilon_n)\right\}\geqslant P\left\{\bigcup_{n=1}^{\infty}(X_n>\varepsilon_n+\delta_n)\right\}\inf_{n\geqslant 1}P\{Y_n\geqslant-\delta_n\};$$

$$P\left\{\overline{\lim_{n\to\infty}}(X_n+Y_n)\geqslant\varepsilon\right\}\geqslant P\left\{\overline{\lim_{n\to\infty}}X_n>\varepsilon+\delta\right\}\underline{\lim_{n\to\infty}}P\{Y_n\geqslant-\delta\}.$$

Proof. Put $A_n = \{X_n > \varepsilon_n + \delta_n\}$, $B_n = \{Y_n \geqslant -\delta_n\}$. By 1.5.c, for $m \geqslant 1$

$$P\left\{\bigcup_{n=m}^{\infty}(X_n + Y_n > \varepsilon_n)\right\} \geqslant P\left\{\bigcup_{n=m}^{\infty} A_n B_n\right\} \geqslant P\left\{\bigcup_{n=m}^{\infty} A_n\right\} \inf_{n \geqslant m} P(B_n).$$

Letting $m = 1$ yields the first inequality while letting $m \to \infty$ and $\varepsilon_n \equiv \varepsilon, \delta_n \equiv \delta$ yields the second one.

The following inequalities all are related to the sum $S_n = \sum_{j=1}^{n} X_j$.

5.3 Symmetrization Inequalities

Let X and X' be i.i.d. r.v.'s, $X^s = X - X'$, mX be the median of X, i.e. the number satisfying $P(X \geqslant mX) \geqslant \frac{1}{2} \leqslant P(X \leqslant mX)$.

5.3.a. (weak symmetrization inequalities). For any x and a,

$$\frac{1}{2}P(X - mX \geqslant x) \leqslant P(X^s \geqslant x);$$

$$\frac{1}{2}P(|X - mX| \geqslant x) \leqslant P(|X^s| \geqslant x) \leqslant 2P(|X - a| \geqslant x/2).$$

Proof.

$$\begin{aligned}
P(X^s \geqslant x) &= P\{(X - mX) - (X' - mX') \geqslant x\} \\
&\geqslant P\{X - mX \geqslant x, X' - mX' \leqslant 0\} \\
&= P(X - mX \geqslant x)P(X' - mX' \leqslant 0) \\
&\geqslant \frac{1}{2}P(X - mX \geqslant x).
\end{aligned}$$

This proves the first inequality, which, together with the inequality obtained by changing X into $-X$, prove the left inequality in the second one. The RHS inequality follows from

$$\begin{aligned}
P(|X^s| \geqslant x) &= P\{|(X - a) - (X' - a)| \geqslant x\} \\
&\leqslant P\left(|X - a| \geqslant \frac{x}{2}\right) + P\left(|X' - a| \geqslant \frac{x}{2}\right) \\
&= 2P\left(|X - a| \geqslant \frac{x}{2}\right).
\end{aligned}$$

5.3.b (symmetrization inequalities). Let $\{X_n, n \geqslant 1\}$ be a sequence of r.v.'s. Then for any $x > 0$ and any sequence $\{c_n, n \geqslant 1\}$ of numbers,

$$\frac{1}{2}P\left\{\sup_{n\geqslant 1}(X_n - mX_n) \geqslant x\right\} \leqslant P\left\{\sup_{n\geqslant 1} X_n^s \geqslant x\right\};$$

$$\frac{1}{2}P\left\{\sup_{n\geqslant 1}|X_n - mX_n| \geqslant x\right\} \leqslant P\left\{\sup_{n\geqslant 1} |X_n^s| \geqslant x\right\}$$

$$\leqslant 2P\left\{\sup_{n\geqslant 1}|X_n - c_n| \geqslant x/2\right\}.$$

Proof. Let $X_n^s = X_n - X_n'$. Putting the events

$$A_n = \{X_n - mX_n \geqslant x\}, \quad B_n = \{X_n' - mX_n' \leqslant 0\}, \quad C_n = \{X_n^s \geqslant x\},$$

we have $A_n B_n \subset C_n$. Using 1.5.c, with $\inf_n P(B_n) \geqslant \frac{1}{2}$, we obtain the first inequality. The second one follows by arguments similar to those used in the proof of 5.3.a.

5.4 Lévy Inequality

Let $X_1, \cdots, X_n$ be independent r.v.'s, $x > 0$.

5.4.a. $P\left\{\max\limits_{1\leqslant j\leqslant n}(S_j - m(S_j - S_n)) \geqslant x\right\} \leqslant 2P(S_n \geqslant x).$

5.4.b. $P\left\{\max\limits_{1\leqslant j\leqslant n}|S_j - m(S_j - S_n)| \geqslant x\right\} \leqslant 2P(|S_n| \geqslant x).$

Proof. Let $S_0 = 0$, $S_k^* = \max\limits_{1\leqslant j\leqslant k}(S_j - m(S_j - S_n))$ and

$$A_k = \{S_{k-1}^* < x, \ S_k - m(S_k - S_n) \geqslant x\},$$

$$B_k = \{S_n - S_k - m(S_n - S_k) \geqslant 0\}.$$

Note that $m(S_n - S_k) = -m(S_k - S_n)$. Then

$$P(S_n \geqslant x) \geqslant P\left(\bigcup_{k=1}^n A_k B_k\right) = \sum_{k=1}^n P(A_k B_k)$$

$$= \sum_{k=1}^n P(A_k)P(B_k) \geqslant \frac{1}{2}\sum_{k=1}^n P(A_k) = \frac{1}{2}P(S_n^* \geqslant x),$$

from which 5.4.a follows. Replacing X_j by $-X_j, 1 \leqslant j \leqslant n$, in 5.4.a we obtain 5.4.b.

The most applicable form of Lévy inequality in limiting theorems is the following corollary.

5.4.c (corollary to Lévy's inequality). Suppose that $EX_j = 0$, $EX_j^2 < \infty$, $j = 1, \cdots, n$. Put $B_n = \sum_{j=1}^{n} EX_j^2$. Then

$$P\left\{\max_{1\leqslant j\leqslant n} S_j \geqslant x\right\} \leqslant 2P(S_n \geqslant x - \sqrt{2B_n}). \tag{16}$$

Proof. By Chebyshev inequality (see 6.1.c),

$$P(|S_j - S_n| \leqslant \sqrt{2ES_n^2}) \geqslant \frac{1}{2} \quad \text{for} \quad j \leqslant n,$$

which implies $|m(S_j - S_n)| \leqslant \sqrt{2ES_n^2}$ and therefore 5.4.c follows from 5.4.a.

Remark. As mentioned in 6.1.e, the inequality $|m(S_j - S_n)| \leqslant \sqrt{2ES_n^2}$ can be improved to $|m(S_j - S_n)| \leqslant \sqrt{ES_n^2}$. Thus, (16) can be sharpened to

$$P\left\{\max_{1\leqslant j\leqslant n} S_j \geqslant x\right\} \leqslant 2P(S_n \geqslant x - \sqrt{B_n}).$$

This inequality is frequently used in proving the strong law of large numbers.

5.5 Bickel Inequality

Let $X_1, \cdots, X_n$ be independent symmetric r.v.'s, and $c_1 \geqslant c_2 \geqslant \cdots \geqslant c_n \geqslant 0$ be constants and let $g(x)$ be a nonnegative convex function[1]. Define

$$G_k = \sum_{j=1}^{k-1}(c_j - c_{j+1})g(S_j) + c_k g(S_k), \quad k = 1, \cdots, n.$$

Then for any $x > 0$,

$$P\left\{\max_{1\leqslant j\leqslant n} c_j g(S_j) \geqslant x\right\} \leqslant 2P(G_n \geqslant x).$$

Proof. At first, we prove that for any $0 \leqslant r \leqslant n-1$ and any real a,

$$P\left\{\sum_{j=1}^{n-r}(c_{r+j} - c_{r+j+1})g(S_j + a) - c_{r+1}g(a) < 0\right\} \leqslant \frac{1}{2}, \quad c_{n+1} = 0. \tag{17}$$

[1] A finite real function g on an interval $J \subset R^1$ is called convex on J if whenever $x_1, x_2 \in J$ and $\lambda \in [0,1]$, $g(\lambda x_1 + (1-\lambda)x_2) \leqslant \lambda g(x_1) + (1-\lambda)g(x_2)$

Recall the following facts for a convex function g: A convex function is continuous and possesses right and left derivatives at every point. Let these derivatives be g'_+ and g'_- respectively. Then for all $x, y, g(x+y) - g(x) \geqslant yg'_\pm(x)$ and $g'_+(x) \geqslant g'_-(x)$, implying that $yg'_\pm(x+y) \geqslant yg'_\pm(x)$. Hence, we can write

$$\begin{aligned}
&\sum_{j=1}^{n-r}(c_{r+j} - c_{r+j+1})g(S_j + a) - c_{r+1}g(a) \\
=&\sum_{j=1}^{n-r}(c_{r+j} - c_{r+j+1})(g(S_j + a) - g(a)) \\
\geqslant&\sum_{j=1}^{n-r} g'_\pm(a)(c_{n+j} - c_{n+j+1})S_j.
\end{aligned}$$

Consequently,

$$\begin{aligned}
&P\left\{\sum_{j=1}^{n-r}(c_{r+j} - c_{r+j+1})g(S_j + a) - c_{r+1}g(a) < 0\right\} \\
\leqslant &P\left\{\sum_{j=1}^{n-r} g'_\pm(a)(c_{n+j} - c_{n+j+1})S_j < 0\right\}
\end{aligned}$$

which proves (17) by symmetry of the r.v.'s.

Let $T = \min\{k \leqslant n : G_k \geqslant x\}$ and $T = n+1$ if no such k exists. Then, since $g \geqslant 0$, we have $G_k \geqslant c_k g(S_k)$. Hence,

$$\begin{aligned}
&P\left\{\max_{1\leqslant j\leqslant n} c_j g(S_j) \geqslant x\right\} \leqslant P\left\{\max_{1\leqslant j\leqslant n} G_j \geqslant x\right\} \\
=&\sum_{k=1}^{n}(P\{T = k, G_n - G_T \geqslant 0\} + P\{T = k, G_n - G_T < 0\}) \\
\leqslant &P(G_n \geqslant x) + \sum_{k=1}^{n} P\{T = k, G_n - G_T < 0\}.
\end{aligned}$$

But

$$\begin{aligned}
&P\{T = k, G_n - G_T < 0\} = \int_{\{T=k\}} P\{G_n - G_k < 0 | X_1, \cdots, X_k\} \mathrm{d}P \\
\leqslant &\int_{\{T=k\}} P\left\{\sum_{j=k+1}^{n}(c_j - c_{j+1})g(S_j) - c_{k+1}g(S_k) < 0 | X_1, \cdots, X_k\right\} \mathrm{d}P \\
\leqslant &\frac{1}{2}P(T = k)
\end{aligned}$$

by using the independence of $X_j's$ and (17). Then Bickel's inequality follows from the above two inequalities.

5.6 Upper Bounds of Tail Probabilities of Partial Sums

Let $X_1, \cdots, X_n$ be independent symmetric r.v.'s. Then for any $1 \leqslant p < 2$, there exists a constant $C(p) > 0$ such that

$$\sup_{x>0} x^p P(|S_n| \geqslant x) \leqslant C(p) \sum_{j=1}^{n} \sup_{x>0} x^p P(|X_j| \geqslant x). \tag{18}$$

Proof. If $\sum_{j=1}^{n} \sup_{x>0} x^p P(|X_j| \geqslant x) = 0$ or ∞, the inequality (18) is trivial. Otherwise, if $\sum_{j=1}^{n} \sup_{x>0} x^p P(|X_j| \geqslant x) = a^p \in (0, \infty)$, we may prove (18) for $Y_j = X_j/a$ and $y = x/a$ instead of x_j and x respectively. Thus, without loss of generality, we may assume that $\sum_{j=1}^{n} \sup_{x>0} x^p P(|X_j| \geqslant x) = 1$. Then for any $x > 0$

$$x^p P(|S_n| \geqslant x) \leqslant 1 + x^p P\left\{\left|\sum_{j=1}^{n} X_j I(|X_j| < x)\right| > x\right\}$$

$$\leqslant 1 + x^{p-2} E\left(\sum_{j=1}^{n} X_j I(|X_j| < x)\right)^2 \leqslant 1 + x^{p-2} \sum_{j=1}^{n} X_j^2 I(|X_j| < x)$$

$$\leqslant 1 + x^{p-2} \int_0^x \sum_{j=1}^{n} P\{|X_j| \geqslant y\} \mathrm{d}y^2 \leqslant 1 + x^{p-2} \int_0^x y^{-p} \mathrm{d}y^2$$

$$= 1 + \frac{2}{2-p} = C(p).$$

The conclusion follows.

5.7 Lower Bounds of Tail Probabilities of Partial Sums

5.7.a. For any r.v.'s $X_1, \cdots, X_n$,

$$P\{\max_{1 \leqslant j \leqslant n} |S_j| \geqslant x\} \geqslant P\{\max_{1 \leqslant j \leqslant n} |X_j| \geqslant 2x\}.$$

Proof. Note that $X_j = S_j - S_{j-1}$.

5.7.b. For independent symmetric r.v.'s $X_1, \cdots, X_n$,

$$2P\{|S_n| \geqslant x\} \geqslant P\{\max_{1 \leqslant j \leqslant n} |X_j| \geqslant x\}.$$

Proof. Let $\tau = \inf\{j : |X_j| \geqslant x\}$. We have

$$P\{|S_n| \geqslant x\} = \sum_{j=1}^{n} P\{|S_n| \geqslant x, \tau = j\}.$$

Now, since for every $j = 1, \cdots, n$, $(-X_1, \cdots, -X_{j-1}, X_j, -X_{j+1}, \cdots, -X_n)$ has the same distribution as $(X_1, \cdots, X_n)$, and $\{\tau = j\}$ only depends on $|X_1|, \cdots, |X_j|$, we also have

$$P\{|S_n| \geqslant x\} = \sum_{j=1}^{n} P\{|X_j - T_j| \geqslant x, \tau = j\},$$

where $T_j = S_n - X_j, j \leqslant n$. Then, summing the two preceding probabilities and noting that $|S_n| + |X_j - T_j| = |X_j + T_j| + |X_j - T_j| \geqslant 2|X_j|$, we obtain

$$2P(|S_n| \geqslant x) \geqslant \sum_{j=1}^{n} P(\tau = j) = P\{\max_{1 \leqslant j \leqslant n} |X_j| \geqslant x\}.$$

5.8 Tail Probabilities for Maximum Partial Sums

Let $X_1, \cdots, X_n$ be independent r.v.'s.

5.8.a (Ottaviani's inequality).

$$P\{\max_{1 \leqslant j \leqslant n} |S_j| \geqslant 2x\} \leqslant \frac{P(|S_n| \geqslant x)}{\min_{1 \leqslant j \leqslant n} P(|S_n - S_j| \leqslant x)}.$$

In particular, if for every $j = 1, \cdots, n$, $P(|S_n - S_j| \leqslant x) \geqslant \frac{1}{2}$, then

$$P\{\max_{1 \leqslant j \leqslant n} |S_j| \geqslant 2x\} \leqslant 2P(|S_n| \geqslant x).$$

Proof. Let $T = \inf\{j : |S_j| \geqslant 2x\}$. Then, by independence, we obtain

$$\begin{aligned} P(|S_n| \geqslant x) &\geqslant P\left\{\bigcup_{j=1}^{n} (T = j, |S_n - S_j| < x)\right\} \\ &= \sum_{j=1}^{n} P(T = j) P(|S_n - S_j| < x), \end{aligned}$$

which implies the desired inequality.

5.8.b (Lévy-Skorohod inequality). For any $0 < c < 1$,

$$P\{\max_{1\leqslant j\leqslant n} S_j \geqslant x\} \leqslant \frac{P(S_n \geqslant cx)}{\min_{1\leqslant j\leqslant n} P(S_n - S_j \geqslant -(1-c)x)}.$$

The proof is similar to that of Ottaviani's inequality.

5.9 Tail Probabilities for Maximum Partial Sums (Continuation)

Let $X_1,\cdots,X_n$ be r.v.'s. Put $S_0=0, S_j=\sum_{k=1}^{j} X_k, M_n=\max_{1\leqslant j\leqslant n}|S_j|, M_n' = \max_{0\leqslant j\leqslant n}(|S_j|\wedge|S_n-S_j|), m_{ijk}=|S_j-S_i|\wedge|S_k-S_j|, L_n=\max_{1\leqslant i\leqslant j\leqslant k\leqslant n} m_{ijk}$.

5.9.a.

$$P(M_n \geqslant x) \leqslant P(M_n' \geqslant x/2) + P(|S_n| \geqslant x/2);$$

$$P(M_n \geqslant x) \leqslant P(M_n' \geqslant x/4) + P\big(\max_{1\leqslant j\leqslant n}|X_j| \geqslant x/4\big).$$

Proof. The first inequality is due to

$$|S_j| \leqslant \min\{|S_n|+|S_j|, |S_n|+|S_n-S_j|\} = |S_n| + |S_j|\wedge|S_n-S_j|. \quad (19)$$

For the second inequality, we only need to show that

$$M_n \leqslant 3M_n' + \max_{1\leqslant j\leqslant n}|X_j|. \quad (20)$$

Consider the set $I=\{j, 0\leqslant j\leqslant n; |S_j|\leqslant|S_n-S_j|\}$. Obviously, $0\in I$. If $S_n = 0$, then $M_n = M_n'$, and (20) trivially holds. We need only consider the case $S_n \neq 0$. In this case, $n \notin I$ and thus there is an integer $j, 0 < j \leqslant n$ such that $j-1\in I$ and $j\notin I$. Consequently, we have $|S_{j-1}| \leqslant |S_n - S_{j-1}|$, $|S_{j-1}| \leqslant M_n'$. Also, $|S_n-S_j| < |S_j|$ and $|S_n-S_j|\leqslant M_n'$. For this j we have

$$|S_n| \leqslant |S_{j-1}| + |X_j| + |S_n - S_j| \leqslant 2M_n' + |X_j|.$$

Then, (20) follows from this and (19).

5.9.b. Suppose that there exist $\gamma \geqslant 0, \alpha > 1$ and nonnegative numbers $u_1,\cdots,u_n$ such that for any $x>0$,

$$P(|S_j - S_i| \geqslant x) \leqslant \left(\sum_{i<l\leqslant j} u_l\right)^{\alpha} \Big/ x^{\gamma}, \qquad 0\leqslant i<j\leqslant n.$$

Then there exists a constant $K_{\gamma,\alpha}$ depending only on γ and α such that

$$P(M_n \geqslant x) \leqslant K_{\gamma,\alpha} \left(\sum_{l=1}^{n} u_l \right)^{\alpha} \Big/ x^{\gamma}.$$

5.9.c. Suppose that there exist $\gamma \geqslant 0, \alpha > 1$ and nonnegative numbers $u_1, \cdots, u_n$ such that for any $x > 0$,

$$P(m_{ijk} \geqslant x) \leqslant \left(\sum_{i<l\leqslant k} u_l \right)^{\alpha} \Big/ x^{2\gamma}, \qquad 0 \leqslant i < j < k \leqslant n.$$

Then there exists constant $K'_{\gamma,\alpha}$ depending only on γ and α such that

$$P(L_n \geqslant x) \leqslant K'_{\gamma,\alpha} \left(\sum_{l=1}^{n} u_l \right)^{\alpha} \Big/ x^{2\gamma}.$$

The proofs of 5.9.b and 5.9.c can be found in Billingsley (1999) (Chapter 2, Section 10).

5.10 Reflection Inequality of Tail Probability (Hoffmann-Jør-gensen)

Let $X_1, \cdots, X_n$ be independent r.v.'s. Then for any $s, t > 0$,

$$P\left\{ \max_{1\leqslant j\leqslant n} |S_j| \geqslant 3t + s \right\} \leqslant \left(P\left\{ \max_{1\leqslant j\leqslant n} |S_j| \geqslant t \right\} \right)^2 + P\left\{ \max_{1\leqslant j\leqslant n} |X_j| \geqslant s \right\}.$$

If the r.v.'s are symmetric, then for any $s, t > 0$,

$$P\{|S_n| \geqslant 2t + s\} \leqslant 4(P\{|S_n| \geqslant t\})^2 + P\left\{ \max_{1\leqslant j\leqslant n} |X_j| \geqslant s \right\}.$$

Proof. Let $\tau = \inf\{k \leqslant n : |S_k| \geqslant t\}$. Then, on $\{\tau = k\}$, $|S_j| \leqslant t$ for $j < k$ and

$$|S_j| \leqslant t + |X_k| + |S_j - S_k|, \quad \text{for} \quad j \geqslant k.$$

Therefore, in either case,

$$\max_{1\leqslant j\leqslant n} |S_j| \leqslant t + \max_{1\leqslant k\leqslant n} |X_k| + \max_{k\leqslant j\leqslant n} |S_j - S_k|.$$

Hence, by independence,

$$\begin{aligned} &P\{\tau = k, \max_{1\leqslant j\leqslant n} |S_j| \geqslant 3t + s\} \\ \leqslant &P\{\tau = k, \max_{1\leqslant j\leqslant n} |X_j| \geqslant s\} + P\{\tau = k\} P\{\max_{k\leqslant j\leqslant n} |S_j - S_k| \geqslant 2t\}. \end{aligned}$$

Since $\max_{k\leqslant j\leqslant n}|S_j - S_k| \leqslant 2\max_{1\leqslant j\leqslant n}|S_j|$, a summation over $k = 1,\cdots,n$ yields the first inequality.

For the second one,

$$|S_n| \leqslant |S_{k-1}| + |X_k| + |S_n - S_k|$$

for each $k = 1,\cdots,n$, so that

$$P\{\tau = k, |S_n| \geqslant 2t + s\} \leqslant P\{\tau = k, \max_{1\leqslant j\leqslant n}|X_j| \geqslant s\} \\ +P\{\tau = k\}P\{|S_n - S_k| \geqslant t\}.$$

Noting $\sum_{k=1}^{n} P\{\tau = k\} = P\Big\{\max_{1\leqslant k\leqslant n}|S_k| \geqslant t\Big\}$, using Lévy's inequality 5.4.b and summing over k yield the second inequality.

5.11 Probability of Maximal Increment (Shao)

Let $\{X_n\}$ be independent r.v.'s. Suppose that there exist $\varepsilon > 0, 0 < \alpha < 1$, and integer $p \geqslant 1$ such that for a certain $x > 0$,

$$P\Big\{\max_{1\leqslant k\leqslant p}|S_k| \geqslant \varepsilon x\Big\} \leqslant \alpha.$$

Then

$$P\left\{\bigcup_{n=0}^{p}(\max_{1\leqslant k\leqslant N}|S_{n+k} - S_n| \leqslant x)\right\} \leqslant \frac{1}{1-\alpha}P\left\{\max_{1\leqslant k\leqslant N}|S_k| \leqslant (1+\varepsilon)x\right\}.$$

Proof. Let

$$E_p = \Big\{\max_{1\leqslant k\leqslant N}|S_{p+k} - S_p| \leqslant x\Big\},$$

$$E_i = \bigcap_{i<n\leqslant p}\Big\{\max_{1\leqslant k\leqslant N}|S_{n+k} - S_N| > x\Big\}\bigcap\Big\{\max_{1\leqslant k<N}|S_{i+k} - S_i| \leqslant x\Big\},$$

$$i = p-1, p-2, \cdots, 0.$$

Apparently,

$$\bigcup_{n=0}^{p}\{\max_{1\leqslant k\leqslant N}|S_{n+k} - S_n| \leqslant x\}$$
$$= \bigcup_{n=0}^{p} E_n \bigcap\{\max_{1\leqslant k\leqslant N}|S_k| < (1+\varepsilon)x\}$$
$$\bigcup\left(\bigcup_{n=1}^{p}\left(E_n\bigcap\{\max_{1\leqslant k\leqslant N}|S_k| \geqslant (1+\varepsilon)x\}\right)\right)$$

$$\subset \{\max_{1\leqslant k\leqslant N}|S_k| < (1+\varepsilon)x\} \bigcup \left(\bigcup_{n=1}^{p}\left(E_n \bigcap \{\max_{1\leqslant k\leqslant n}|S_k| \geqslant (1+\varepsilon)x\}\right)\right)$$

$$\bigcup \left(\bigcup_{n=1}^{p}\left(E_n \bigcap \{\max_{n\leqslant k\leqslant N}|S_k| \geqslant (1+\varepsilon)x\}\right)\right)$$

$$\subset \{\max_{1\leqslant k\leqslant N}|S_k| < (1+\varepsilon)x\} \bigcup \left(\bigcup_{n=1}^{p}\left(E_n \bigcap \left\{\max_{1\leqslant k\leqslant n}|S_k| \geqslant (1+\varepsilon)x\right\}\right)\right)$$

$$\bigcup \left(\bigcup_{n=1}^{p}\left(E_n \bigcap \{|S_n| \geqslant \varepsilon x\}\right)\right)$$

$$\subset \{\max_{1\leqslant k\leqslant N}|S_k| < (1+\varepsilon)x\} \bigcup \left(\bigcup_{n=1}^{p}\left(E_n \bigcap \{\max_{1\leqslant k\leqslant n}|S_k| \geqslant \varepsilon x\}\right)\right).$$

Noting that E_n and $\{\max_{1\leqslant k\leqslant n}|S_k| \geqslant \varepsilon x\}$ are independent, we have

$$P\left\{\bigcup_{n=0}^{p}\left(\max_{1\leqslant k\leqslant N}|S_{n+k}-S_n| \leqslant x\right)\right\}$$

$$\leqslant P\{\max_{1\leqslant k\leqslant N}|S_k| \leqslant (1+\varepsilon)x\} + \sum_{n=1}^{p} P(E_n)P\{\max_{1\leqslant k\leqslant n}|S_k| \geqslant \varepsilon x\}$$

$$\leqslant P\{\max_{1\leqslant k\leqslant N}|S_k| \leqslant (1+\varepsilon)x\} + \alpha\sum_{n=1}^{p} P(E_n)$$

$$\leqslant P\{\max_{1\leqslant k\leqslant N}|S_k| \leqslant (1+\varepsilon)x\} + \alpha P\left\{\bigcup_{n=0}^{p}(\max_{1\leqslant k\leqslant N}|S_{n+k}-S_n| \leqslant x)\right\}.$$

This proves the desired inequality.

5.12 Mogulskii Minimal Inequality

Let $X_1, \cdots, X_n$ be independent r.v.'s, $2 \leqslant m \leqslant n$, $x_1, x_2 > 0$. Then

$$P\{\min_{m\leqslant k\leqslant n}|S_k| \leqslant x_1\} \leqslant P\{|S_n| \leqslant x_1+x_2\}/\min_{m\leqslant k\leqslant n} P\{|S_m - S_k| \leqslant x_2\}.$$

Proof.

$$P\{|S_n| \leqslant x_1+x_2\} \geqslant \sum_{k=m}^{n} P\{\min_{m\leqslant j\leqslant k-1}|S_j| > x_1, |S_k| \leqslant x_1, |S_n| \leqslant x_1+x_2\}$$

$$\geqslant \sum_{k=m}^{n} P\{\min_{m\leqslant j\leqslant k-1} |S_j| > x_1, |S_k| \leqslant x_1\} P\{|S_n - S_k| \leqslant x_2\}$$
$$\geqslant P\{\min_{m\leqslant k\leqslant n} |S_k| \leqslant x_1\} \min_{m\leqslant k\leqslant n} P\{|S_n - S_k| \leqslant x_2\}.$$

5.13 Wilks Inequality

Let X have a distribution function $F(x_1, \cdots, x_k)$ and let $F(x_i)$, $i = 1, \cdots, k$, denote the marginal distributions. We have

$$F(x_1, \cdots, x_k) \leqslant \left(\prod_{i=1}^{k} F(x_i)\right)^{1/k}.$$

Proof. By the Hölder inequality, for any random variables $Y_1, \cdots, Y_k$, we have $E|Y_1 \cdots Y_k| \leqslant \left(E|Y_1|^k \cdots E|Y_k|^k\right)^{1/k}$. Then, the Wilks inequality follows by taking $Y_i = I(X_i \leqslant x_i)$, $i = 1, \cdots, k$.

References

Billingsley P. 1999. Convergence of Probability Measures. 2nd ed.. New York: J. Wiley.

Bickel P J. 1970. A Hájek-Rényi extension of Lévy inequality and some applications. Acta Math. Acad. Sci. Hungar., 21: 199-206.

Chow Y S, Teicher H. 1997. Probability Theory: Independence, Interchangeability, Martingales. 2nd ed., New York: Springer-Verlag.

Hall P, Heyde C C. 1980. Martingale Limit Theory and Its Application. New York: Academic Press.

Hoffmann-Jørgensen J. 1974. Sums of independent Banach space valued random variables. Studia Math., 52: 159-186.

Loéve M. 1977. Probability Theory. 4th ed.. New York: Springer-Verlag.

Mogulskii A A. 1980. On the law of the iterated logarithm in Chung's form for functional spaces. Theor. Probab. Appl., 24: 405-413.

Petrov V V. 1995. Limit Theorems of Probability Theory: Sequences of Independent Random Variables. Oxford & New York: Clarendon Press & Oxford University Press.

Shao Q M. 1992. How small are the increments of partial sums of independent random variables. Scientia Sinica, 35(A): 675-689.

Shorack G R, Wellner J A. 1986. Empirical processes with applications to statistics. New York: Wiley.

Wilks Samuel S. 1962. Mathematical Statistics. A Wiley Publication in Mathematical Statistics. New York-London: John Wiley & Sons, Inc..

Chapter 6

Bounds of Probabilities in Terms of Moments

Let X be a r.v. with the d.f. $F(x)$. If $\int |x| \mathrm{d}F(x) < \infty$, we say that the (mathematical) expectation, or mean, of X exists. The expectation is defined by $\int x \mathrm{d}F(x)$ and denoted by EX.

Let k be a positive number. If EX^k and $E|X|^k$ exist, then we call them the moment and the absolute moment of order k (about the origin) respectively. If $E(X-EX)^k$ and $E|X-EX|^k$ exist, they are called the central moment and the absolute central moment of order k, respectively. In particular, $E(X-EX)^2$ is called the variance of X, usually denoted by $\mathrm{Var}X$.

Let $(\Omega, \mathscr{F}, P)$ be a probability space, X be a r.v. defined on Ω with $E|X| \leqslant \infty$, $\mathscr{A}$ be a sub-σ-algebra of $\mathscr{F}$. The conditional expectation of X given $\mathscr{A}$, denoted by $E(X|\mathscr{A})$, is defined to be a r.v. satisfying

(i) $E(X|\mathscr{A})$ is $\mathscr{A}$-measurable,

(ii) $\int_A E(X|\mathscr{A}) \mathrm{d}P = \int_A X \mathrm{d}P$ for all $A \in \mathscr{A}$.

Let $\{Y_n, n \geqslant 1\}$ be a sequence of r.v.'s with $E|Y_n| < \infty$, $\{\mathscr{A}_n, n \geqslant 1\}$ be an increasing sequence of sub-σ-algebras for which Y_n is $\mathscr{A}_n$-measurable (we say that Y_n is adapted to $\mathscr{A}_n$.) The sequence $\{Y_n, \mathscr{A}_n, n \geqslant 1\}$ is called a martingale, if for every n,

$$E(Y_{n+1}|\mathscr{A}_n) = Y_n \qquad \text{a.s.}$$

It is called a submartingale (or supermartingale) if the "=" in the above equality is replaced by "$\geqslant$" (correspondingly "$\leqslant$"). If $\{Y_n, \mathscr{A}_n\}$ is a martingale, putting $X_n = Y_n - Y_{n-1}$ with $Y_0 = 0$, $\{X_n, n \geqslant 1\}$ is called a martingale difference sequence.

6.1 Chebyshev-Markov Type Inequalities

6.1.a (general form). Let X be a r.v. and $g(x) > 0$ be a nondecreasing function on R. Then for any x,

$$P(X \geqslant x) \leqslant \frac{Eg(X)}{g(x)}. \tag{21}$$

Proof.

$$P(X \geqslant x) \leqslant \int_{\{g(X)\geqslant g(x)\}} \mathrm{d}P \leqslant \frac{Eg(X)I(g(X) \geqslant g(x))}{g(x)} \leqslant \frac{Eg(X)}{g(x)}.$$

Here, and in the sequel, $I(\cdot)$ denotes the indicator function of the set in the braces, that is, it takes value 1 or 0 according to whether or not the sample point belongs to the set.

6.1.b (inequality with lower bounds). Let $g(x) > 0$ be an even function and nondecreasing on $[0, \infty)$. Suppose that $Eg(X) < \infty$. Then for any $x > 0$,

$$\frac{Eg(X) - g(x)}{\text{a.s.}\sup g(X)} \leqslant P(|X| \geqslant x) \leqslant \frac{Eg(X)}{g(x)}. \tag{22}$$

in which $\text{a.s.}\sup g(X) = \inf\{t : P(g(X) > t) = 0\}$.

Proof. The right part is the same as (21), and the left follows from

$$\begin{aligned} Eg(X) &= \int_{\{|X|\geqslant x\}} g(X)\mathrm{d}P + \int_{\{|X|<x\}} g(X)\mathrm{d}P \\ &\leqslant \text{a.s.}\sup g(X)P(|X| \geqslant x) + g(x). \end{aligned}$$

Two important consequences of general form are the following inequalities:

6.1.c (Chebyshev inequality). For any $x > 0$,

$$P(|X - EX| \geqslant x) \leqslant \mathrm{Var}(X)/x^2.$$

6.1.d (Markov inequality). For any $r > 0$ and $x > 0$,

$$P(|X| \geqslant x) \leqslant E|X|^r/x^r.$$

6.1.e (generalization of the Chebyshev inequality). Put $\sigma^2 = \mathrm{Var}(X)$. For any x and a,

$$P(X - EX \geqslant x) \leqslant \frac{\sigma^2 + a^2}{(x + a)^2}.$$

By taking $a = \sigma^2/x$,

$$P(X - EX \geqslant x) \leqslant \frac{\sigma^2}{x^2 + \sigma^2}.$$

Furthermore, letting $x = \sigma$, we have $P(X \geqslant EX + \sigma) \leqslant 1/2$, which implies $m(X) \leqslant EX + \sigma$. By symmetry, we obtain $|EX - m(X)| \leqslant \sigma$.
Proof. $P(X - EX \geqslant x) \leqslant \frac{E(X-EX+a)^2}{(x+a)^2} = \frac{\sigma^2+a^2}{(x+a)^2}$.
6.1.f (discrete case). Let X be a discrete r.v. with possible values $1, 2, \cdots$. Suppose that $P(X = k)$ is nonincreasing in k. Then for all $k \geqslant 1$,

$$P(X = k) \leqslant \frac{2}{k^2} EX.$$

Proof.

$$\begin{aligned} EX &= \sum_{j=1}^{\infty} jP(X = j) \geqslant \sum_{j=1}^{k} jP(X = j) \geqslant \sum_{j=1}^{k} jP(X = k) \\ &= \frac{k(k+1)}{2} P(X = k) > \frac{k^2}{2} P(X = k). \end{aligned}$$

6.2 Lower Bounds

6.2.a. If $|X| \leqslant 1$, then

$$P(|X| \geqslant x) \geqslant EX^2 - x^2.$$

This is a special case of the left hand side (LHS) inequality in 6.1.b.
6.2.b. If $X \geqslant 0$, then for any $0 < x < 1$,

$$P(X > xEX) \geqslant (1 - x)^2 (EX)^2 / EX^2.$$

Proof. By the Cauchy-Schwarz inequality (see 8.4.b),

$$\begin{aligned} EX &= EXI(|X| > xEX) + EXI(|X| \leqslant xEX) \\ &\leqslant (EX^2 P(X > xEX))^{1/2} + xEX. \end{aligned}$$

6.3 Series of Tail Probabilities

If $X \geqslant 0$, then

$$\sum_{n=1}^{\infty} P(X \geqslant n) \leqslant EX \leqslant \sum_{n=0}^{\infty} P(X \geqslant n).$$

Proof. The inequalities follow from the following observation

$$EX = \int_0^\infty x\mathrm{d}P(X < x) = \int_0^\infty P(X \geqslant x)\mathrm{d}x = \sum_{n=0}^\infty \int_0^1 P(X \geqslant n+x)\mathrm{d}x.$$

6.4 Kolmogorov Type Inequalities

Let $X_1, \cdots, X_n$ be independent r.v.'s with $EX_j = 0, x > 0$.

6.4.a (Kolmogorov inequality).

$$P\left\{\max_{1\leqslant j\leqslant n} |S_j| \geqslant x\right\} \leqslant \mathrm{Var}(S_n)/x^2.$$

If, in addition, there is a constant $c > 0$ such that $|X_j| \leqslant c, 1 \leqslant j \leqslant n$, we also have

$$P\left\{\max_{1\leqslant j\leqslant n} |S_j| \geqslant x\right\} \geqslant 1 - \frac{(x+c)^2}{\mathrm{Var}(S_n)}.$$

Proof. Let $S_0 = 0, A_k = \{\max_{1\leqslant j<k} |S_j| < x \leqslant |S_k|\}$. Noting independence of $S_k I(A_k)$ and $S_n - S_k$, we have

$$\begin{aligned}\int_{A_k} S_n^2 \mathrm{d}P &= \int_{A_k} (S_k + S_n - S_k)^2 \mathrm{d}P = \int_{A_k} S_k^2 \mathrm{d}P + \int_{A_k} (S_n - S_k)^2 \mathrm{d}P \\ &\geqslant \int_{A_k} S_k^2 \mathrm{d}P \geqslant x^2 P(A_k).\end{aligned}$$

Summing over $k = 1, \cdots, n$, we obtain

$$\mathrm{Var}(S_n) \geqslant x^2 P\left(\bigcup_{k=1}^n A_k\right) = x^2 P\left\{\max_{1\leqslant j\leqslant n} |S_j| \geqslant x\right\}.$$

The upper bound part is proved.

Consider the case of $|X_j| \leqslant c$. Let $B_0 = \Omega, B_k = \{\max_{1\leqslant j\leqslant k} |S_j| < x\}$. Since

$$S_{k-1} I(B_{k-1}) + X_k I(B_{k-1}) = S_k I(B_{k-1}) = S_k I(B_k) + S_k I(A_k)$$

and $S_{k-1} I(B_{k-1})$ and X_k are independent while $I(B_k) I(A_k) = 0$, it follows that

$$E(S_{k-1} I(B_{k-1}))^2 + EX_k^2 P(B_{k-1}) = E(S_k I(B_k))^2 + E(S_k I(A_k))^2.$$

Since $P(B_{k-1}) \geqslant P(B_n)$ and $|X_k| \leqslant c$, and hence

$$|S_kI(A_k)| \leqslant |S_{k-1}I(A_k)| + |X_kI(A_k)| \leqslant (x+c)I(A_k),$$

it follows that

$$E(S_{k-1}I(B_{k-1}))^2 + EX_k^2P(B_n) \leqslant E(S_kI(B_k))^2 + (x+c)^2P(A_k).$$

Summing up over $k = 1, \cdots, n$ we obtain

$$\begin{aligned}\left(\sum_{k=1}^n EX_k^2\right) P(B_n) &\leqslant E(S_nI(B_n))^2 + (x+c)^2P\left(\bigcup_{k=1}^n A_k\right)\\ &\leqslant x^2P(B_n) + (x+c)^2P(B_n^c) \leqslant (x+c)^2,\end{aligned}$$

which implies the second inequality.

6.4.b (generalized Kolmogorov inequality). Let $r \geqslant 1$. Put $A = \{\max_{1\leqslant j\leqslant n} |S_j| \geqslant x\}$. We have

$$x^rP\{\max_{1\leqslant j\leqslant n} |S_j| \geqslant x\} \leqslant E|S_n|^rI(A) \leqslant E|S_n|^r.$$

Proof. Let $S_0 = 0, A_k = \{\max_{1\leqslant j\leqslant k} |S_j| < x \leqslant |S_k|\}$. By the moment inequality in section 8.7,

$$E|S_n|^rI(A) = \sum_{k=1}^n E|S_n|^rI(A_k) \geqslant \sum_{k=1}^n E|S_k|^rI(A_k) \geqslant x^rP(A).$$

6.4.c (another generalization).

$$P\{\max_{1\leqslant j\leqslant n} S_j \geqslant x\} \leqslant \mathrm{Var}(S_n)/(x^2 + \mathrm{Var}(S_n)).$$

Proof. For any $a > 0$,

$$\begin{aligned}\int_{A_k} (S_n+a)^2\mathrm{d}P &= \int_{A_k} (S_k + a + S_n - S_k)^2\mathrm{d}P\\ &= \int_{A_k} (S_k+a)^2\mathrm{d}P + \int_{A_k} (S_n - S_k)^2\mathrm{d}P\\ &\geqslant \int_{A_k} (S_k+a)^2\mathrm{d}P \geqslant (x+a)^2P(A_k).\end{aligned}$$

From this it follows that

$$P\{\max_{1\leqslant j\leqslant n} S_j \geqslant x\} \leqslant [\mathrm{Var}(S_n) + a^2]/(x+a)^2.$$

The generalized Kolmogorov inequality follows by choosing $a = \mathrm{Var}(S_n)/x$.

6.5 Generalization of Kolmogorov Inequality for a Submartingale

Let $\{Y_n, \mathscr{A}_n, n \geqslant 1\}$ be a submartingale, $x > 0$.

6.5.a (Doob inequality).

$$xP\{\max_{1\leqslant j\leqslant n} Y_j \geqslant x\} \leqslant \int_{\{\max\limits_{1\leqslant j\leqslant n} Y_j \geqslant x\}} Y_n \mathrm{d}P \leqslant EY_n^+ \leqslant E|Y_n|.$$

Proof. Let $\alpha = \inf\{k \leqslant n : Y_k \geqslant x\}$, and $\alpha = n+1$ if no such k exists. Clearly Y_α is a r.v., and $\{\alpha = k\} \in \mathscr{A}_k$. Moreover by the definitions of a submartingale and conditional expectation, for $k < n$,

$$\int_{\{\alpha=k\}} Y_k \mathrm{d}P \leqslant \int_{\{\alpha=k\}} E(Y_{k+1}|\mathscr{A}_k)\mathrm{d}P = \int_{\{\alpha=k\}} Y_{k+1}\mathrm{d}P$$

and inductively

$$\int_{\{\alpha=k\}} Y_k \mathrm{d}P \leqslant \int_{\{\alpha=k\}} Y_n \mathrm{d}P.$$

Then

$$\begin{aligned} xP\{\max_{1\leqslant j\leqslant n} Y_j \geqslant x\} &\leqslant \int_{\{\max\limits_{1\leqslant j\leqslant n} Y_j \geqslant x\}} Y_\alpha \mathrm{d}P \\ &\leqslant \sum_{k=1}^{n} \int_{\{\max\limits_{1\leqslant j\leqslant n} Y_j \geqslant x, \alpha=k\}} Y_k \mathrm{d}P = \sum_{k=1}^{n} \int_{\{\alpha=k\}} Y_k \mathrm{d}P \\ &\leqslant \sum_{k=1}^{n} \int_{\{\alpha=k\}} Y_n \mathrm{d}P = \int_{\{\max\limits_{1\leqslant k\leqslant n} Y_k \geqslant x\}} Y_n \mathrm{d}P. \end{aligned}$$

6.5.b.

$$xP\{\min_{1\leqslant j\leqslant n} Y_j \leqslant -x\} \leqslant E(Y_n - Y_1) - \int_{\{\min\limits_{1\leqslant j\leqslant n} Y_j \leqslant -x\}} Y_n \mathrm{d}P \leqslant EY_n^+ - EY_1.$$

Proof. Let $\beta = \inf\{k \leqslant n : Y_k \leqslant -x\}$, and $\beta = n+1$ if no such k exists. Put

$$A_k = \{\min_{1\leqslant j\leqslant k} Y_j \leqslant -x\} = \{\beta \leqslant k\}.$$

β is a stopping time relative to $\{\mathscr{A}_n\}$. Hence $\{Y_1, Y_{\beta\wedge n}\}$ is a two-term submartingale (cf. Chung (1974), Theorem 9.3.4), and so

$$\begin{aligned} EY_1 \leqslant EY_\beta &= \int_{\{\beta\leqslant n\}} Y_\beta \mathrm{d}P + \int_{A_n^c} Y_n \mathrm{d}P \\ &\leqslant -xP(A_n) + EY_n - \int_{A_n} Y_n \mathrm{d}P. \end{aligned}$$

6.5.c. Let $1 \leqslant m \leqslant n, A_m \in \mathscr{A}_m$ and $A = \{\max_{1\leqslant j\leqslant n} Y_j \geqslant x\}$, then

$$xP(A_m A) \leqslant \int_{A_m A} Y_n \mathrm{d}P.$$

Proof. Same as for 6.5.a.

6.5.d. If $\{Y_n, \mathscr{A}_n, n \geqslant 1\}$ is a martingale, $p \geqslant 1$, then for any $x > 0$,

$$x^p P\{\max_{1\leqslant j\leqslant n} |Y_j| \geqslant x\} \leqslant \int_{\{\max_{1\leqslant j\leqslant n} Y_j \geqslant x\}} |Y_n|^p \mathrm{d}P \leqslant E|Y_n|^p.$$

Proof. Apply 6.5.a to the submartingale $\{|Y_n|^p\}$.

6.6 Rényi-Hájek Type Inequalities

This is a generalization of the Kolmogorov inequality for the weighted sums. We give a result on a submartingale. A consequence in the independent case is clear.

6.6.a (Rényi-Hájek-Chow inequality). Let $\{Y_n, \mathscr{A}_n, n \geqslant 1\}$ be a submartingale, r.v. $\tau_j \in \mathscr{A}_{j-1}$ with $\tau_1 \geqslant \tau_2 \geqslant \cdots \geqslant \tau_n > \tau_{n+1} = 0$ a.s. Then for any $x > 0, 1 \leqslant m \leqslant n$,

$$P\{\max_{m\leqslant j\leqslant n} \tau_j Y_j \geqslant x\} \leqslant \{E(\tau_m Y_m^+) + \sum_{j=m+1}^{n} E(\tau_j(Y_j^+ - Y_{j-1}^+))\}/x.$$

Proof. Let $A_k = \{\max_{m\leqslant j<k} \tau_j|Y_j| < x \leqslant \tau_k|Y_k|\}, m \leqslant k \leqslant n$. Similar to the proof in 6.5.a, for $k \leqslant j$ we have

$$\int_{A_k} \tau_{j+1} Y_j \mathrm{d}P \leqslant \int_{A_k} \tau_{j+1} Y_{j+1} \mathrm{d}P.$$

Hence

$$\begin{aligned}\int_{A_k} \tau_k Y_k \mathrm{d}P &= \int_{A_k} (\tau_k - \tau_{k+1}) Y_k \mathrm{d}P + \int_{A_k} \tau_{k+1} Y_k \mathrm{d}P \\ &\leqslant \int_{A_k} (\tau_k - \tau_{k+1}) Y_k \mathrm{d}P + \int_{A_k} \tau_{k+1} Y_{k+1} \mathrm{d}P.\end{aligned}$$

Repeating this procedure we obtain

$$\int_{A_k} \tau_k Y_k \mathrm{d}P \leqslant \int_{A_k} \sum_{j=k}^{n} (\tau_j - \tau_{j+1}) Y_j \mathrm{d}P.$$

Then

$$
\begin{aligned}
xP\{\max_{m\leqslant j\leqslant n}\tau_jY_j\geqslant x\} &= x\sum_{k=m}^{n}P(A_k)\\
&\leqslant\sum_{k=m}^{n}\int_{A_k}\tau_kY_k\mathrm{d}P\leqslant\sum_{k=m}^{n}\sum_{j=k}^{n}\int_{A_k}(\tau_j-\tau_{j+1})Y_j\mathrm{d}P\\
&\leqslant\sum_{j=m}^{n}\sum_{k=m}^{j}\int_{A_k}(\tau_j-\tau_{j+1})Y_j^+\mathrm{d}P\\
&\leqslant\sum_{j=m}^{n}\int_{\{\max_{m\leqslant k\leqslant j}\tau_kY_k\geqslant x\}}(\tau_j-\tau_{j+1})Y_j^+\mathrm{d}P\\
&\leqslant\sum_{j=m}^{n}E(\tau_j-\tau_{j+1})Y_j^+\\
&=E(\tau_mY_m^+)+\sum_{j=m+1}^{n}E(\tau_j(Y_j^+-Y_{j-1}^+)).
\end{aligned}
$$

6.6.b (a special case). Let $\{X_n,n\geqslant 1\}$ be a martingale difference sequence with $\sigma_n^2\equiv EX_n^2<\infty, n=1,2,\cdots$, let constants $c_1\geqslant c_2\geqslant\cdots\geqslant c_n>0, 1\leqslant m\leqslant n$. Put $S_n=\sum_{j=1}^{n}X_j$. Then for any $x>0$,

$$
P\{\max_{m\leqslant j\leqslant n}c_j|S_j|\geqslant x\}\leqslant\frac{1}{x^2}\Big(c_m^2\sum_{j=1}^{m}\sigma_j^2+\sum_{j=m+1}^{n}c_j^2\sigma_j^2\Big).
$$

6.6.c (further generalization). Let $Y_1,Y_2,\cdots,Y_n$ be r.v.'s and $\mathscr{A}_1,\mathscr{A}_2,\cdots,\mathscr{A}_n$ be σ-algebras. Suppose that Y_j is adaptive to $\mathscr{A}_j$ $(j=1,2,\cdots,n)$, and

$$
E(|Y_j||\mathscr{A}_{j-1})\geqslant a_j|Y_{j-1}|\quad\text{a.s.}\quad\text{for}\quad 1\leqslant j\leqslant n
$$

(where $Y_0=0$) with $0\leqslant a_j\leqslant 1$ for every j. Let constants $c_1\geqslant c_2\geqslant\cdots\geqslant c_n>c_{n+1}=0, r\geqslant 1, 1\leqslant m\leqslant n$. Then for any $x>0$,

$$
\begin{aligned}
P\{\max_{m\leqslant j\leqslant n}c_j|Y_j|\geqslant x\}&\leqslant\sum_{j=m}^{n}(c_j^r-a_{j+1}^rc_{j+1}^r)E|Y_j|^r/x^r\\
&=\sum_{j=m}^{n}c_j^r(E|Y_j|^r-a_j^rE|Y_{j-1}|^r)/x^r.
\end{aligned}
$$

Proof. Along the lines of the proof of 6.6.a and note that

$$c_j = \sum_{k=j}^{n}(c_k - a_{k+1}c_{k+1})\left(\prod_{i=j+1}^{k} a_i\right), \qquad \text{here } \prod_{i=k+1}^{k} a_i = 1.$$

6.6.d (without moment condition). Let $\{Y_1, \cdots, Y_n\}$ be a martingale difference sequence with $EY_j = 0, EY_j^2 < \infty, j = 1, \cdots, n$ and $\{Z_1, \cdots, Z_n\}$ be a sequence of random variables. Write $X_j = Y_j + Z_j$, and suppose that $c_1 \geqslant c_2 \geqslant \cdots \geqslant c_n > 0$, $1 \leqslant m \leqslant n$. Put $S_n = \sum_{j=1}^{n} X_j$. Then for any $x > 0$ and $0 < \varepsilon < 1$,

$$P\{\max_{m\leqslant j\leqslant n} c_j|S_j| \geqslant x\} \leqslant \frac{1}{(1-\varepsilon)^2x^2}\left(c_m^2\sum_{j=1}^{m} EY_j^2 + \sum_{j=m+1}^{n} c_j^2 EY_j^2\right)$$
$$+2\sum_{j=m+1}^{n} P(Z_j \neq 0) + P\left\{c_n\left|\sum_{j=1}^{m} Z_j\right| \geqslant \frac{1}{2}\varepsilon x\right\}.$$

Proof. Put $U_n = \sum_{j=1}^{n} Y_j, V_n = \sum_{j=1}^{n} Z_j$. Then

$$P\{\max_{m\leqslant j\leqslant n} c_j|S_j| \geqslant x\} \leqslant P\left\{\bigcup_{j=m}^{n}(c_j|U_j| \geqslant (1-\varepsilon)x)\right\}$$
$$+P\left\{\bigcup_{j=m}^{n}(c_j|V_j| \geqslant \varepsilon x)\right\}.$$

For the first term on the right hand side, we have, by 6.6.b,

$$P\left\{\bigcup_{j=m}^{n}(c_j|U_j| \geqslant (1-\varepsilon)x)\right\} = P\{\max_{m\leqslant j\leqslant n} c_j|U_j| \geqslant (1-\varepsilon)x\}$$
$$\leqslant \frac{1}{(1-\varepsilon)^2x^2}\left(c_m^2\sum_{j=1}^{m} EY_j^2 + \sum_{j=m+1}^{n} c_j^2 EY_j^2\right).$$

To deal with the second term, set $A_j = \{c_j|V_j| \geqslant \varepsilon x\}$. Then

$$P\left\{\bigcup_{j=m}^{n}(c_j|V_j| \geqslant \varepsilon x)\right\} = P(A_n) + \sum_{j=m}^{n-1} P\left\{\bigcap_{k=j+1}^{n} A_k^c \bigcap A_j\right\},$$

while

$$P\left\{\bigcap_{k=j+1}^{n} A_k^c \bigcap A_j\right\} \leqslant P(Z_{j+1} \neq 0)$$

and

$$\begin{aligned} P(A_n) &\leqslant P\left\{c_n|V_m| \geqslant \frac{1}{2}\varepsilon x\right\} + P\left\{c_n|\sum_{j=m+1}^{n} Z_j| \geqslant \frac{1}{2}\varepsilon x\right\} \\ &\leqslant P\left\{c_n|V_m| \geqslant \frac{1}{2}\varepsilon x\right\} + \sum_{j=m+1}^{n} P(Z_j \neq 0). \end{aligned}$$

Combining these inequalities, we conclude the proof of the desired result.

6.7 Chernoff Inequality

Let $\{X_n, n \geqslant 1\}$ be a sequence of i.i.d. r.v.'s with $EX_1 = 0$ and the moment generating function (m.g.f.) $M(t) = Ee^{tX_1}$. Let

$$m(x) = \inf_t e^{-xt} M(t).$$

Then for any $x > 0$,

$$P\{S_n \geqslant nx\} \leqslant m(x)^n$$

and

$$\lim_{n\to\infty} P^{1/n}\{S_n \geqslant nx\} = m(x). \tag{23}$$

Proof. The first inequality is from

$$P\{S_n \geqslant nx\} \leqslant \inf_t e^{-tnx} Ee^{tS_n} = (\inf_t e^{-xt} M(t))^n = m(x)^n$$

by 6.1.a.

Now, we prove the second conclusion. If for some $x_0 > 0$, $P(X_1 \geqslant x_0) = 0$, then for all $x \geqslant x_0$, $P(S_n \geqslant nx) = 0$ and $m(x) = 0$. Then (23) is trivially true. We assume that $P(X_1 > x) > 0$ for all x. Then, X_1 is not degenerate and it is easy to see that $R_x(t) \equiv e^{-xt}M(t)$ attains its minimum at a finite t and $m(x) > 0$. Let $\tau \equiv \tau(x) = \inf\{t : m(x) = R_x(t)\}$. Write $F(y) = P(X_1 - x < y)$ and define

$$G(z) = \int_{-\infty}^{z} e^{\tau y} dF(y)/m(x).$$

It is a d.f. of some r.v., say Z. Let $\xi(t)$ denote the m.g.f. of Z. Then

$$\xi(t) = \int \mathrm{e}^{tu}\mathrm{d}G(u) = R_x(\tau + t)/m(x).$$

Hence $EZ = \xi'(t)|_{t=0} = m'(\tau)/m(x) = 0$ by the definition of τ. Furthermore $\sigma^2 \equiv \mathrm{Var}Z > 0$ since X_1 is non-degenerate. Let $Z_1, Z_2, \cdots$ be a sequence of iid. r.v.'s with the common d.f. $G(\cdot)$. Put

$$U_n = \frac{1}{\sigma\sqrt{n}}\sum_{j=1}^{n} Z_j \qquad \text{and} \qquad H_n(x) = P(U_n < x).$$

By the central limit theorem, $\lim_{n\to\infty} H_n(x) = \Phi(x)$. Then for any $\varepsilon > 0$,

$$\begin{aligned}
P\{S_n \geqslant nx\} &= P\left\{\sum_{j=1}^{n}(X_j - x) \geqslant 0\right\} \\
&= \int_{y_1+\cdots+y_n\geqslant 0}\cdots\int \mathrm{d}F(y_1)\cdots\mathrm{d}F(y_n) \\
&= m(x)^n \int_{z_1+\cdots+z_n\geqslant 0}\cdots\int \mathrm{e}^{-\tau(z_1+\cdots+z_n)}\mathrm{d}G(z_1)\cdots\mathrm{d}G(z_n) \\
&= m(x)^n \int_0^\infty \mathrm{e}^{-\sqrt{n}\sigma\tau u}\mathrm{d}H_n(u) \\
&= m(x)^n\sqrt{n}\sigma\tau \int_0^\infty \mathrm{e}^{-\sqrt{n}\sigma\tau u}(H_n(u) - H_n(0))\mathrm{d}u \\
&\geqslant m(x)^n\sqrt{n}\sigma\tau \int_\varepsilon^\infty \mathrm{e}^{-\sqrt{n}\sigma\tau u}(H_n(u) - H_n(0))\mathrm{d}u \\
&\geqslant m(x)^n(H_n(\varepsilon) - H_n(0))\sqrt{n}\sigma\tau \int_\varepsilon^\infty \mathrm{e}^{-\sqrt{n}\sigma\tau u}\mathrm{d}u \\
&= m(x)^n(H_n(\varepsilon) - H_n(0))\mathrm{e}^{-\sqrt{n}\sigma\tau\varepsilon},
\end{aligned}$$

which implies that

$$\liminf_{n\to\infty}(n^{-1/2}\log(m(x)^{-n}P\{S_n \geqslant nx\})) \geqslant -\sigma\tau\varepsilon.$$

By the arbitrariness of ε, it follows that $n^{-1/2}\log(m(x)^{-n}P\{S_n \geqslant nx\}) = o(1)$, which implies

$$\frac{1}{n}\log P\{S_n \geqslant nx\} = \log m(x) + o(1),$$

proving the second conclusion.

Remark. From the proof, one can see that there is a more precise result which is the existence of positive numbers $b_1, b_2, \cdots$, with $\log b_n = o(1)$, such that

$$P\{S_n \geqslant nx\} = \frac{b_n}{(2\pi n)^{1/2}} m(x)^n (1 + o(1))$$

(cf. Bahadur and Rao (1960)).

6.8 Fuk and Nagaev Inequality

Let $X_j, j = 1, \cdots, n$, be independent r.v.'s with d.f.'s $F_j(x), j = 1, \cdots, n$. Let $x, y_1, \cdots, y_n$ be positive numbers. Put $\eta = (y_1, \cdots, y_n), y = \max(y_1, \cdots, y_n)$,

$$A(r, \eta) = \sum_{j=1}^{n} \int_{|u| \leqslant y_j} |u|^r \mathrm{d}F_j(u),$$

where $0 < r \leqslant 1$. Then

$$P(S_n \geqslant x) \leqslant \sum_{j=1}^{n} P(|X_j| \geqslant y_j) + \exp\left\{\frac{x}{y} - \frac{x}{y} \log\left(\frac{xy^{r-1}}{A(r, \eta)} + 1\right)\right\}.$$

If $xy^{r-1} > A(r, \eta)$, then

$$P(|S_n| \geqslant x) \leqslant \sum_{j=1}^{n} P(|X_j| \geqslant y_j) + \exp\left\{\frac{x}{y} - \frac{A(r, \eta)}{y^r} - \frac{x}{y} \log\left(\frac{xy^{r-1}}{A(r, \eta)}\right)\right\}.$$

Proof. Let

$$\bar{X}_j = \begin{cases} X_j, & \text{if } |X_j| \leqslant y_j, \\ 0, & \text{otherwise}, \end{cases} \qquad j = 1, \cdots, n,$$

$$\bar{S}_n = \sum_{j=1}^{n} \bar{X}_j.$$

Clearly, for any $t > 0$,

$$\begin{aligned} P(|S_n| \geqslant x) &\leqslant P(\bar{S}_n \neq S_n) + P(|\bar{S}_n| \geqslant x) \\ &\leqslant \sum_{j=1}^{n} P(|X_j| \geqslant y_j) + \mathrm{e}^{-tx} E\mathrm{e}^{t|\bar{S}_n|}. \end{aligned} \tag{24}$$

We estimate $Ee^{t|\bar{S}_n|}$ as follows. Observe that $|u|^{-1}(e^{t|u|}-1)$ attains its maximum value in the region $|u| \leqslant z$ for $|u| = z$. Therefore

$$\begin{aligned} Ee^{t|\bar{S}_n|} &\leqslant \prod_{j=1}^{n} Ee^{t|\bar{X}_j|} = \prod_{j=1}^{n} E\left(1 + \frac{e^{t|\bar{X}_j|}-1}{|\bar{X}_j|}|\bar{X}_j|\right) \\ &\leqslant \prod_{j=1}^{n}\left(1 + \frac{e^{ty_j}-1}{y_j}\int_{|u|\leqslant y_j}|u|dF_j(u)\right) \\ &\leqslant \prod_{j=1}^{n}\left(1 + \frac{e^{ty}-1}{y^r}\int_{|u|\leqslant y_j}|u|^r dF_j(u)\right) \\ &\leqslant \exp\left(\frac{e^{ty}-1}{y^r}A(r,\eta)\right). \end{aligned}$$

Hence

$$e^{-tx}Ee^{t|\bar{S}_n|} \leqslant \exp\left(\frac{e^{ty}-1}{y^r}A(r,\eta) - tx\right). \tag{25}$$

Then, the first inequality follows by substituting (25) into (24) with

$$t = \frac{1}{y}\log\left(\frac{xy^{r-1}}{A(r,\eta)} + 1\right).$$

The second inequality follows by minimizing (25) with respect to t and noticing the condition $xy^{r-1} > A(r,\eta)$ which implies the minimizer $t = y^{-1}\log(xy^{r-1}A^{-1}(r,\eta)) > 0$.

6.9 Burkholder Inequality

Let $\{Y_n, \mathscr{A}_n\}$ be a martingale or nonnegative submartingale. Then for any $x > 0$,

$$P\Big\{\sum_{n=1}^{\infty}(Y_n - Y_{n-1})^2 \geqslant x\Big\} \leqslant 3\sup_{n\geqslant 1} E|Y_n|/\sqrt{x}.$$

Proof. We only need consider the case where $\sup\limits_{n\geqslant 1} E|Y_n| < \infty$. Under this condition, by the martingale convergence theorem, there exists a r.v. Y_∞ such that $Y_n \to Y_\infty$ a.s. and $E|Y_\infty| \leqslant \sup_{n\geqslant 1} E|Y_n|$. Set $Y_0 = 0$, $X_n = Y_n - Y_{n-1}$, $\mu = \inf\{n \geqslant 1 : |Y_n| > \sqrt{x}\}$ with the convention that $\mu = \infty$ if no such n exists, hence for this case Y_μ are $Y_{\mu-1}$ and well defined. By the dominated convergence theorem, we have

$$EY_\mu Y_{\mu-1}I(\mu=\infty)=EY_\infty^2\leqslant\sqrt{x}\lim_{n\to\infty}E|Y_n|I(\mu=\infty).$$

On the other hand, since $|Y_{\mu-1}|\leqslant\sqrt{x}$, we have

$$\begin{aligned}EY_\mu Y_{\mu-1}I(\mu<\infty)&=\liminf_{n\to\infty}\sum_{k=1}^n\sqrt{x}E|Y_kI(\mu=k)|\\ &\leqslant\sqrt{x}\liminf_{n\to\infty}\sum_{k=1}^n E|Y_n|I(\mu=k)\leqslant\sqrt{x}\liminf_{n\to\infty}E|Y_n|I(\mu\leqslant n)\\ &\leqslant\sqrt{x}\liminf_{n\to\infty}E|Y_n|I(\mu<\infty).\end{aligned}$$

Combining the two inequalities, we obtain

$$EY_\mu Y_{\mu-1}\leqslant\sqrt{x}\sup_{n\geqslant1}E|Y_n|.$$

Let $\nu=\mu\wedge n$. By the same lines we can prove that

$$EY_\nu Y_{\nu-1}\leqslant\sqrt{x}\sup_{n\geqslant1}E|Y_n|.$$

Furthermore

$$Y_{n-1}^2=\sum_{j=1}^{n-1}X_j^2+2\sum_{j=1}^{n-1}Y_{j-1}X_j.$$

Thus

$$\begin{aligned}\sum_{j=1}^{n-1}X_j^2+Y_{n-1}^2&=2Y_{n-1}^2+2Y_{n-1}X_n-2\sum_{j=1}^nY_{j-1}X_j\\ &=2Y_nY_{n-1}-2\sum_{j=1}^nY_{j-1}X_j.\end{aligned}$$

Clearly, the above equality holds with n replaced by ν. Hence, by noticing $Y_{j-1}\geqslant0$ for the submartingale case, we obtain

$$E\sum_{j=1}^{\nu}Y_{j-1}X_j=\sum_{j=1}^{n}E\{I(\mu\geqslant j)Y_{j-1}E(X_j|\mathscr{A}_{j-1})\}\geqslant0.$$

Thus

$$E\sum_{j=1}^{\nu-1}X_j^2\leqslant2EY_\nu Y_{\nu-1}\leqslant2\sqrt{x}\sup_{n\geqslant1}E|Y_n|.$$

Making $n \to \infty$, one gets

$$E \sum_{j=1}^{\mu-1} X_j^2 \leqslant 2\sqrt{x} \sup_{n \geqslant 1} E|Y_n|.$$

Since $\sum_{j=1}^{\mu-1} X_j^2 = \sum_{j=1}^{\infty} X_j^2$ on $\{\mu = \infty\} = \{\sup_{n \geqslant 1} E|Y_n| \leqslant \sqrt{x}\}$, we have

$$P\left\{\sum_{n=1}^{\infty} X_n^2 \geqslant x, \sup_{n \geqslant 1} |Y_n| \leqslant \sqrt{x}\right\} \leqslant P\left\{\sum_{j=1}^{\mu-1} X_j^2 \geqslant x\right\}$$

$$\leqslant E \sum_{j=1}^{\mu-1} X_j^2 / x \leqslant 2 \sup_{n \geqslant 1} E|Y_n| / \sqrt{x}.$$

Therefore

$$P\left\{\sum_{n=1}^{\infty} X_n^2 \geqslant x\right\} \leqslant P\{\sup_{n \geqslant 1} |Y_n| > \sqrt{x}\} + P\left\{\sum_{n=1}^{\infty} X_n^2 \geqslant x, \sup_{n \geqslant 1} |Y_n| \leqslant \sqrt{x}\right\}$$

$$\leqslant 3 \sup_{n \geqslant 1} E|Y_n| / \sqrt{x}.$$

6.10 Complete Convergence of Partial Sums

Let $\{X_n, n \geqslant 1\}$ be i.i.d. r.v.'s, $EX_1 = 0$. Then there exist positive constants C_1 and C_2 such that for any $x > 0$,

$$C_1 x^{-2} E X_1^2 I(|X_1| \geqslant x) \leqslant \sum_{n=1}^{\infty} P(|S_n| \geqslant xn) \leqslant C_2 x^{-2} E X_1^2 I(|X_1| \geqslant x).$$

See the proof in (Pruss, 1997).

References

Bahadur R R, Rao R R. 1960. On deviations of the sample mean. Ann. Math. Statist., 31: 1015-1027.

Bickel P J. 1970. A Hájek-Rényi extension of Lévy inequality and some applications. Acta Math. Acad. Sci. Hungar., 21: 199-206.

Birnbaum Z, Marshall A. 1961. Some multivariate Chebyshev inequalities with extensions to continuous parameter processes. Ann. Math. Statist., 32: 687-703.

Burkholder D L. 1973. The 1971 Wald memorial lectures: distribution function inequalities for martingales. Ann. Probab., 1: 19-42.

Chernoff H. 1952. A measure of asymptotic efficiency for tests of a hypothesis based on the sum of observations. Ann. Math. Statist., 23: 493-507.

Chow Y S, Teicher H. 1997. Probability Theory: Independence, Interchangeability, Martingales. 2nd ed., New York: Springer-Verlag.

Chung K L. 1974. A Course in Probability Theory. 2nd ed.. New York: Academic Press.

Doob J L. 1953. Stochastic Processes. New York: Wiley.

Fuk D H, Nagaev S V. 1971. Probabilistic inequalities for sums of independent random variables. Theor. Probab. Appl., 16: 643-660.

Hájek J., Rényi A. 1955. Generalization of an inequality of Kolmogorov. Acta Math. Acad. Sci. Hungar., 6: 281-283.

Hall P, Heyde C C. 1980. Martingale Limit Theory and Its Application. New York: Academic Press.

Heyde C C. 1968. An extension of the Hájek-Rényi inequality for the case without moment condition. J. Appl. Probab., 5: 481-483.

Petrov V V. 1995. Limit Theorems of Probability Theory: Sequences of Independent Random Variables. Oxford & New York: Clarendon Press & Oxford University Press.

Pruss A R. 1997. A two-side estimate in the Hsu-Robbins-Erdös law of large numbers. Stoch. Proc. Appl., 70: 173-180.

Shorack G R, Wellner J A. 1986. Empirical processes with applications to statistics. New York: Wiley.

Stout W. 1974. Almost Sure Convergence. New York: Academic Press.

Chapter 7

Exponential Type Estimates of Probabilities

Exponential rate is the best in convergence in probability. Such inequalities are important when investigating the law of large numbers and the law of iterated logarithm. Some of such inequalities are well known and frequently employed in statistics and probability, such as Hoeffding, Bernstein, Bennett and Kolmogorov inequalities. These inequalities can be found in most textbooks on limiting theorems, such as Loève (1977), Petrov (1995). Some new inequalities will be referenced therein.

7.1 Equivalence of Exponential Estimates

7.1.a. There exist positive constants b and c such that

$$P(|X| \geqslant x) \leqslant b\mathrm{e}^{-cx} \qquad \text{for all} \quad x > 0.$$

7.1.b. There exist a constant $H > 0$ such that

$$E\mathrm{e}^{tX} < \infty \qquad \text{for} \quad |t| < H.$$

7.1.c. There exist a constant $a > 0$ such that

$$E\mathrm{e}^{a|X|} < \infty.$$

7.1.d. There exist positive constants g and T such that

$$E\mathrm{e}^{t(X-EX)} \leqslant \mathrm{e}^{gt^2} \qquad \text{for} \quad |t| \leqslant T.$$

Proof. Denote the d.f. of X by $F(x)$. Note that

$$E\mathrm{e}^{a|X|} = \int_{-\infty}^{0} \mathrm{e}^{-ax}\mathrm{d}F(x) + \int_{0}^{\infty} \mathrm{e}^{ax}\mathrm{d}F(x) \leqslant E\mathrm{e}^{-aX} + E\mathrm{e}^{aX}.$$

Hence 7.1.b and 7.1.c are equivalent.

If 7.1.c is true, then

$$P(|X| \geqslant x) \leqslant \mathrm{e}^{-ax} E\mathrm{e}^{a|X|}$$

for every $x > 0$ by 6.1.a. We can conclude that 7.1.a is also true. We shall now show that 7.1.a implies 7.1.c. In fact, from 7.1.a, we have

$$F(x) \leqslant b\mathrm{e}^{-c|x|} \quad \text{for} \quad x \leqslant 0 \quad \text{and} \quad 1 - F(x) \leqslant b\mathrm{e}^{-cx} \quad \text{for} \quad x > 0.$$

Hence for $0 < a < c$, by integration by parts, we obtain

$$\begin{aligned} E\mathrm{e}^{a|X|} &= \int_{-\infty}^{0} \mathrm{e}^{-ax} \mathrm{d}F(x) - \int_{0}^{\infty} \mathrm{e}^{ax} \mathrm{d}(1 - F(x)) \\ &\leqslant 2F(0) + ab \int_{-\infty}^{0} \mathrm{e}^{(c-a)x} \mathrm{d}x + ab \int_{0}^{\infty} \mathrm{e}^{-(c-a)x} \mathrm{d}x < \infty. \end{aligned}$$

At last, we show that 7.1.b and 7.1.d are equivalent. Obviously, 7.1.d implies 7.1.b. Conversely, if 7.1.b is true, then

$$\log E\mathrm{e}^{t(X-EX)} = \frac{t^2}{2} \mathrm{Var}(X) + o(t^2) \qquad \text{as} \quad t \to 0.$$

For any constant $g > \frac{1}{2}\mathrm{Var}(X)$, the inequalities $\log E\mathrm{e}^{t(X-EX)} \leqslant gt^2$ and $E\mathrm{e}^{t(X-EX)} \leqslant \mathrm{e}^{gt^2}$ hold for all sufficiently small t, i.e., 7.1.d is true.

In the following sections 7.2 to 7.8, assume $X_1, \cdots, X_n$ are independent r.v.'s and write $S_n = \sum_{j=1}^{n} X_j$.

7.2 Petrov Exponential Inequalities

7.2.a. Suppose that there exist positive constants $g_1, \cdots, g_n, T$ such that

$$E\mathrm{e}^{tX_j} \leqslant \mathrm{e}^{g_k t^2/2}, \qquad j = 1, \cdots, n,$$

for $0 \leqslant t \leqslant T$ (or for $-T \leqslant t \leqslant 0$). Then, putting $G = \sum_{j=1}^{n} g_j$, we have

$$P(S_n \geqslant x) \leqslant \mathrm{e}^{-x^2/2G} \qquad (\text{correspondingly } P(S_n \leqslant -x) \leqslant \mathrm{e}^{-x^2/2G})$$

for $0 \leqslant x \leqslant GT$ and

$$P(S_n \geqslant x) \leqslant \mathrm{e}^{-Tx/2} \qquad (\text{correspondingly } P(S_n \leqslant -x) \leqslant \mathrm{e}^{-Tx/2})$$

for $x \geqslant GT$.

Proof. Obviously, it is enough to consider the case of $0 < t \leqslant T$. For every x, we have

$$P(S_n \geqslant x) \leqslant \mathrm{e}^{-tx} E\mathrm{e}^{tS_n} = \mathrm{e}^{-tx} \prod_{j=1}^{n} E\mathrm{e}^{tX_j} \leqslant \mathrm{e}^{-tx+Gt^2/2}. \tag{26}$$

For fixed $x, 0 < x \leqslant GT$ (the case of $x = 0$ is obvious), the function $f(t) = \frac{Gt^2}{2} - tx$ attains its minimum at $t = x/G$, which satisfies the condition $0 < t \leqslant T$. Taking $t = x/G$ in (26) yields the first inequality.

Now assume $x \geqslant GT$. At this time, $f(t)$ is non-increasing. Putting $t = T$ in (26) gives the second inequality.

7.2.b. Suppose that $EX_j = 0, j = 1, \cdots, n$, and that there exists a constant $H > 0$ such that

$$|EX_j^m| \leqslant \frac{m!}{2} \sigma_j^2 H^{m-2}, \qquad j = 1, \cdots, n,$$

for all $m \geqslant 2$, where $\sigma_j^2 = EX_j^2$. Put $B_n = \sum\limits_{j=1}^{n} \sigma_j^2$. Then for $0 \leqslant x \leqslant B_n/H$,

$$P(S_n \geqslant x) \leqslant \mathrm{e}^{-x^2/(4B_n)}, \quad P(S_n \leqslant -x) \leqslant \mathrm{e}^{-x^2/(4B_n)}$$

and for $x \geqslant B_n/H$,

$$P(S_n \geqslant x) \leqslant \mathrm{e}^{-Tx/2}, \quad P(S_n \leqslant -x) \leqslant \mathrm{e}^{-Tx/2}.$$

Proof. For $|t| \leqslant 1/(2H)$,

$$\begin{aligned} E\mathrm{e}^{tX_j} &= 1 + \frac{t^2}{2}\sigma_j^2 + \frac{t^3}{6} EX_j^3 + \cdots \\ &\leqslant 1 + \frac{t^2}{2}\sigma_j^2 (1 + H|t| + H^2 t^2 + \cdots) \\ &\leqslant 1 + \frac{t^2 \sigma_j^2}{2(1 - H|t|)} \leqslant 1 + t^2 \sigma_j^2 \leqslant \mathrm{e}^{t^2 \sigma_j^2}. \end{aligned}$$

Applying 7.2.a, we obtain the assertion in 7.2.b.

7.2.c. If $EX_j \geqslant 0, j = 1, \cdots, n$, then the assertion in 7.2.a can be strengthened as follows:

$$P(\max_{1 \leqslant j \leqslant n} S_j \geqslant x) \leqslant \mathrm{e}^{-x^2/(2G)}, \quad \text{if} \quad 0 \leqslant x \leqslant GT,$$

$$P(\max_{1 \leqslant j \leqslant n} S_j \geqslant x) \leqslant \mathrm{e}^{-Tx/2}, \quad \text{if} \quad x \geqslant GT.$$

Proof. We just need to point out the following fact. Since $EX_j \geqslant 0, j = 1, \cdots, n$, $\{S_j, 1 \leqslant j \leqslant n\}$ is a submartingale, and furthermore, so is $\{e^{tS_j}, 1 \leqslant j \leqslant n\}$ for any $t \geqslant 0$. Hence we can use the Doob inequality 6.5.a.

7.3 Hoeffding Inequality

Suppose that $0 \leqslant X_j \leqslant 1$. Put $\mu_j = EX_j$, $\bar{X} = \frac{1}{n}\sum_{j=1}^{n} X_j$, $\mu = E\bar{X}$. Then for $0 < x < 1 - \mu$,

$$\begin{aligned} P(\bar{X} - \mu \geqslant x) &\leqslant P\{\max_{1\leqslant j\leqslant n}(S_j - ES_j) \geqslant nx\} \\ &\leqslant \left\{\left(\frac{\mu}{\mu + x}\right)^{\mu+x}\left(\frac{1-\mu}{1-\mu-x}\right)^{1-\mu-x}\right\}^n \\ &\leqslant e^{-g(\mu)nx^2} \leqslant e^{-2nx^2}, \end{aligned}$$

where

$$g(\mu) = \begin{cases} \dfrac{1}{1-2\mu}\log\dfrac{1-\mu}{\mu}, & \text{for} \quad 0 < \mu < \dfrac{1}{2}, \\ \dfrac{1}{2\mu(1-\mu)}, & \text{for} \quad \dfrac{1}{2} \leqslant \mu < 1. \end{cases}$$

If there exist $a_j \leqslant b_j$, such that $a_j \leqslant X_j \leqslant b_j, j = 1, \cdots, n$, then for any $x > 0$,

$$\begin{aligned} P(\bar{X} - \mu \geqslant x) &\leqslant P\{\max_{1\leqslant j\leqslant n}(S_j - ES_j) \geqslant nx\} \\ &\leqslant \exp\left\{-2n^2x^2 / \sum_{j=1}^{n}(b_j - a_j)^2\right\}. \end{aligned}$$

Proof. We only prove the first inequality as the proof of the second one is similar. Let $t > 0$. Since e^{tx} is a convex function of x, we have

$$e^{tx} \leqslant (1 - x) + xe^t, \quad \text{for} \quad 0 \leqslant x \leqslant 1.$$

Hence the m.g.f $M_j(t)$ of X_j satisfies

$$M_j(t) \leqslant 1 - \mu_j + \mu_j e^t.$$

Thus, noting that the geometric mean does not exceed the arithmetic mean and using the Doob inequality to the submartingale $e^{t(S_j - ES_j)}$, we obtain

$$\begin{aligned}
&P\{\max_{1\leqslant j\leqslant n}(S_j - ES_j) \geqslant nx\}\\
&= P\{\max_{1\leqslant j\leqslant n}\exp\{t(S_j - ES_j)\} \geqslant \exp(nxt)\}\\
&\leqslant \mathrm{e}^{-nxt} E\mathrm{e}^{t(S_n - ES_n)} = \mathrm{e}^{-n(\mu+x)t}\prod_{j=1}^{n} M_j(t)\\
&\leqslant \mathrm{e}^{-n(\mu+x)t}\prod_{j=1}^{n}(1-\mu_j(1-\mathrm{e}^t))\\
&\leqslant \mathrm{e}^{-n(\mu+x)t}\left(\frac{1}{n}\sum_{j=1}^{n}(1-\mu_j(1-\mathrm{e}^t))\right)^n\\
&= (\mathrm{e}^{-(\mu+x)t}(1-\mu(1-\mathrm{e}^t)))^n.
\end{aligned}$$

The RHS is minimized by the choice

$$t_0 = \log\frac{(1-\mu)(\mu+x)}{\mu(1-\mu-x)} > 0.$$

Then we obtain

$$P(\bar{X}-\mu \geqslant x) \leqslant \left\{\left(\frac{\mu}{\mu+x}\right)^{\mu+x}\left(\frac{1-\mu}{1-\mu-x}\right)^{1-\mu-x}\right\}^n \equiv \exp(-nx^2 G(x,\mu)),$$

where

$$G(x,\mu) = \frac{\mu+x}{x^2}\log\frac{\mu+x}{\mu} + \frac{1-\mu-x}{x^2}\log\frac{1-\mu-x}{1-\mu}.$$

We will minimize $G(x,\mu)$ with respect to x for $0 < x < 1-\mu$, and the resulting minimum will be denoted as $g(\mu)$. Now

$$\begin{aligned}
x^2\frac{\partial G(x,\mu)}{\partial x} &= \left(1 - 2\frac{1-\mu}{x}\right)\log\left(1-\frac{x}{1-\mu}\right)\\
&\quad - \left(1-2\frac{\mu+x}{x}\right)\log\left(1-\frac{x}{\mu+x}\right),
\end{aligned} \tag{27}$$

where $0 < x/(1-\mu) < 1$ and $0 < x/(\mu+x) < 1$. We expand the function

$$\begin{aligned}
H(s) &\equiv \left(1-\frac{2}{s}\right)\log(1-s)\\
&= 2 + \left(\frac{2}{3}-\frac{1}{2}\right)s^2 + \left(\frac{2}{4}-\frac{1}{3}\right)s^3 + \left(\frac{2}{5}-\frac{1}{4}\right)s^4 + \cdots
\end{aligned}$$

with all positive coefficients. Thus $H(s)$ is increasing for $0 < s < 1$. Hence we see from (27) that $(\partial/\partial x)G(x,\mu) > 0$ if and only if $x/(1-\mu) > x/(\mu+x)$ or equivalently $x > 1-2\mu$. $G(x,\mu)$ achieves its minimum at $x = 1-2\mu$; while if $1-2\mu \leqslant 0$, then $G(x,\mu)$ achieves its minimum at $x=0$. Inserting these values into $G(x,\mu)$ yields its minimum

$$g(\mu) = \begin{cases} \dfrac{1}{1-2\mu}\log\dfrac{1-\mu}{\mu}, & \text{for } \ 0 \leqslant \mu < \dfrac{1}{2}, \\ \dfrac{1}{2\mu(1-\mu)}, & \text{for } \ \dfrac{1}{2} \leqslant \mu < 1. \end{cases}$$

Moreover, $g(\mu) \geqslant g(\frac{1}{2}) = 2$ obviously. The first inequality can be proved by combining these results.

7.4 Bennett Inequality

Suppose that $X_j \leqslant b, EX_j = 0, j = 1,\cdots,n$. Put $\sigma_j^2 = EX_j^2$ and $\sigma^2 = \frac{1}{n}\sum\limits_{j=1}^{n}\sigma_j^2$. Then for any $x > 0$,

$$\begin{aligned} P(\bar{X} > x) &\leqslant P\{\max_{1\leqslant j\leqslant n} S_j \geqslant nx\} \\ &\leqslant \exp\left\{-\frac{nx}{b}\left[\left(1+\frac{\sigma}{bx}\right)\log\left(1+\frac{bx}{\sigma^2}\right)-1\right]\right\}. \end{aligned}$$

Proof. Also using the Doob inequality, for any $t > 0$ we have

$$P(S_n \geqslant nx) \leqslant P\{\max_{1\leqslant j\leqslant n} S_j \geqslant nx\} \leqslant \mathrm{e}^{-nxt}\prod_{j=1}^{n} E\mathrm{e}^{tX_j}.$$

Noting that $EX_j = 0$ and $g(x) \equiv (\mathrm{e}^x - 1 - x)/x^2$ (conventionally with $g(0) = \frac{1}{2}$) is nonnegative, increasing and convex on R, we have

$$E\mathrm{e}^{tX_j} \leqslant 1 + t^2\sigma_j^2 g(tb) \leqslant \exp\left\{\sigma_j^2\frac{\mathrm{e}^{tb}-1-tb}{b^2}\right\}.$$

Hence

$$P\{\max_{1\leqslant j\leqslant n} S_j \geqslant nx\} \leqslant \exp\left(-nxt + n\sigma^2\frac{\mathrm{e}^{tb}-1-tb}{b^2}\right).$$

Minimizing the exponent by the choice $t = (1/b)\log(1 + bx/\sigma^2)$, we obtain

$$\begin{aligned}
&P\{\max_{1\leqslant j\leqslant n} S_j \geqslant nx\}\\
&\leqslant \exp\left\{-\frac{nx}{b}\log\left(1+\frac{bx}{\sigma^2}\right)+\frac{n\sigma^2}{b^2}\left(\frac{bx}{\sigma^2}-\log\left(1+\frac{bx}{\sigma^2}\right)\right)\right\}\\
&= \exp\left\{-\frac{nx}{b}\left[\left(1+\frac{\sigma^2}{bx}\right)\log\left(1+\frac{bx}{\sigma^2}\right)-1\right]\right\}.
\end{aligned}$$

Remark. If $0 < x < b$, the bound can be changed to

$$\left\{\left(1+\frac{bx}{\sigma^2}\right)^{-(1+bx/\sigma^2)\sigma^2/(b^2+\sigma^2)}\left(1-\frac{x}{b}\right)^{-(1-x/b)b^2/(b^2+\sigma^2)}\right\}^n.$$

7.5 Bernstein Inequality

Suppose that $EX_j = 0$ and $E|X_j|^n \leqslant \sigma_j^2 n! a^{n-2}/2$ for all $n \geqslant 2$ where $\sigma_j^2 = EX_j^2, a > 0$. Then, putting $\sigma^2 = \frac{1}{n}\sum_{j=1}^{n}\sigma_j^2$, for any $x > 0$,

$$P(S_n \geqslant \sqrt{n}x) \leqslant P\left\{\max_{1\leqslant j\leqslant n} S_j \geqslant \sqrt{n}x\right\} \leqslant \exp\left\{-\frac{\sqrt{n}x^2}{2(\sqrt{n}\sigma^2 + ax)}\right\}.$$

Proof. The LHS inequality is trivial. To prove the RHS inequality, we first note that for any $t > 0$, e^{tx} is a convex function and that $\{\mathrm{e}^{tS_j}, j \leqslant n\}$ forms a submartingale. Let $t > 0$ satisfy $ta \leqslant c < 1$. Then, we have

$$P\left\{\max_{1\leqslant j\leqslant n} S_j \geqslant \sqrt{n}x\right\} = P\left\{\max_{1\leqslant j\leqslant n} \exp(tS_j) \geqslant \mathrm{e}^{tx\sqrt{n}}\right\} \leqslant \mathrm{e}^{-tx\sqrt{n}}\prod_{j=1}^{n} E\mathrm{e}^{tX_j},$$

$$\begin{aligned}
E\mathrm{e}^{tX_j} &= 1 + \frac{t^2}{2}EX_j^2 + \frac{t^3}{3!}EX_j^3 + \cdots\\
&\leqslant 1 + \frac{t^2}{2}\sigma_j^2 + \frac{t^3}{2}\sigma_j^2 a + \cdots\\
&\leqslant 1 + \frac{t^2\sigma_j^2}{2(1-c)} \leqslant \exp\left\{\frac{t^2\sigma_j^2}{2(1-c)}\right\},
\end{aligned}$$

Hence

$$\mathrm{e}^{-\sqrt{n}xt} E\mathrm{e}^{tS_n} \leqslant \exp\left\{-\sqrt{n}xt + \frac{t^2 n\sigma^2}{2(1-c)}\right\}. \tag{28}$$

The exponent achieves its minimum at $t_0 = (1-c)x/(\sqrt{n}\sigma^2)$. Letting $t_0 a = c$, we obtain

$$c = \frac{ax}{\sqrt{n}\sigma^2 + ax} < 1,$$

and

$$t_0 = \frac{x}{\sqrt{n}\sigma^2 + ax}.$$

Inserting them into (28) yields the required inequality.

Remark. If $|X_j| \leqslant a$ a.s., then the moment condition is met.

7.6 Exponential Bounds for Sums of Bounded Variables

Suppose $EX_j = 0$ and $|X_j| \leqslant d_j$ a.s., $j = 1, \cdots, n$, where $d_1, \cdots, d_n$ are positive constants. Let $x > 0$. Put $a = \left(\sum_{j=1}^{n} d_j^2\right)^{1/2}$ and $b = \max_{1 \leqslant j \leqslant n} j^{1/p}|d_j|$ for some $1 < p < 2$.

7.6.a. For any $x > 0$, $P(|S_n| \geqslant x) \leqslant 2\exp(-x^2/2a^2)$.

Proof. Note that the function $y \to \exp(ty)$ is convex and $ty = t(1+y)/2 - t(1-y)/2$. For $|y| \leqslant 1$,

$$\exp(ty) \leqslant \cosh t + y \sinh t \leqslant \exp(t^2/2) + y \sinh t.$$

Hence, applying this inequality with $y = X_j/d_j$, we have

$$E\exp(tS_n) \leqslant \prod_{j=1}^{n} \exp(t^2 d_j^2/2) = \exp(t^2a^2/2).$$

By taking $t = x/a^2$, it follows that

$$P(S_n \geqslant x) \leqslant \exp\{-tx + t^2a^2/2\} = \exp(-x^2/2a^2).$$

7.6.b. With $q = p/(p-1)$, there exists a constant $c_q > 0$ such that

$$P(S_n \geqslant x) \leqslant 2\exp(-c_q x^q/b^q).$$

Proof. For any integers $0 < m \leqslant n$ write

$$|S_n| \leqslant \sum_{j=1}^{m} |X_j| + \left| \sum_{j=m+1}^{n} X_j \right| \leqslant b \sum_{j=1}^{m} j^{-1/p} + \left| \sum_{j=m+1}^{n} X_j \right|$$
$$\leqslant bqm^{1/q} + \left| \sum_{j=m+1}^{n} X_j \right|.$$

Consider first the case where $x > 2bq$ and let $m = \max\{j : x > 2bqj^{1/q}\}$. Then by 7.8.a,

$$P(|S_n| \geqslant x) \leqslant P\left(\left| \sum_{j=m+1}^{n} X_j \right| \geqslant bqm^{1/q} \right)$$
$$\leqslant 2\exp\left\{ -b^2q^2m^{2/q} \Big/ \left(2 \sum_{j=m+1}^{n} d_j^2 \right) \right\},$$

where

$$\sum_{j=m+1}^{n} d_j^2 \leqslant b^2 \sum_{j=m+1}^{n} j^{-2/p} \leqslant \frac{b^2 q}{q-2} m^{1-2/p}.$$

Hence

$$P(|S_n| \geqslant x) \leqslant 2\exp\{-q(q-2)m/2\} \leqslant 2\exp(-c_q' x^q / b^q),$$

where $c_q' = q(q-2)/(4(2q)^q)$.

When $x \leqslant 2bq$, taking $c_q'' = (\log 2)/(2q)^q$, we have

$$P(|S_n| \geqslant x) \leqslant 1 \leqslant 2\exp(-c_q'' x^q / b^q).$$

The inequality follows with $c_q = c_q' \vee c_q''$.

7.7 Kolmogorov Inequalities

Suppose that $EX_j = 0$, $\sigma_j^2 \equiv EX_j^2 < \infty$, $|X_j| \leqslant cs_n$ a.s., $j = 1, \cdots, n$, where $c > 0$ is a constant and $s_n^2 = \sum_{j=1}^{n} \sigma_j^2$. Let $x > 0$.

7.7.a (the upper bound). If $xc \leqslant 1$, then

$$P(S_n/s_n \geqslant x) \leqslant \exp\left\{ -\frac{x^2}{2} \left(1 - \frac{xc}{2} \right) \right\};$$

and if $xc \geqslant 1$, then

$$P(S_n/s_n \geqslant x) \leqslant \exp\left\{-\frac{x}{4c}\right\}.$$

7.7.b (the lower bound). For given $\gamma > 0$, there exist $x(\gamma)$ and $\pi(\gamma)$ such that for all $x(\gamma) < x < \pi(\gamma)/c$,

$$P(S_n/s_n \geqslant x) \geqslant \exp\left\{-\frac{x^2}{2}(1+\gamma)\right\}.$$

Proof. Let $t > 0$. It follows that for $tcs_n \leqslant 1$,

$$\begin{aligned} Ee^{tX_j} &\leqslant 1 + \frac{t^2\sigma_j^2}{2}\left(1 + \frac{tcs_n}{3} + \frac{t^2c^2s_n^2}{4\cdot 3} + \cdots\right) \\ &\leqslant \exp\left\{\frac{t^2\sigma_j^2}{2}\left(1 + \frac{tcs_n}{2}\right)\right\}. \end{aligned}$$

Then

$$e^{-ts_nx}Ee^{tS_n} \leqslant \exp\left\{-ts_nx + \frac{t^2s_n^2}{2}\left(1 + \frac{tcs_n}{2}\right)\right\}$$

from which 7.7.a follows if t is chosen to be x/s_n or $1/(cs_n)$ according to $xc \leqslant 1$ or $\geqslant 1$.

We begin to proceed with the proof of 7.7.b. Let α and β be small positive constants to be determined later by the constant γ. Set $t = x/(1-\beta)$. Then, $tc \leqslant 2\alpha < 1$ by choosing $\pi(\gamma)$ to be small enough. Thus,

$$\begin{aligned} \prod_{j=1}^{n} Ee^{tX_j/s_n} &\geqslant \prod_{j=1}^{n}\left(1 + \frac{t^2\sigma_j^2}{2s_n^2}\left(1 - \frac{tc}{3} - \frac{t^2c^2}{4\cdot 3} - \cdots\right)\right) \\ &\geqslant \exp\left\{\frac{t^2}{2}\left(1 - \frac{tc}{2}\right)\right\} \geqslant \exp\left\{\frac{t^2}{2}(1-\alpha)\right\}. \end{aligned} \tag{29}$$

On the other hand, putting $q(y) = P(S_n/s_n \geqslant y)$ and from integration by parts, we have

$$Ee^{tS_n/s_n} = t\int e^{ty}q(y)dy.$$

Split the real line $\mathbb{R}$ into the five parts $I_1 = (-\infty, 0]$, $I_2 = (0, t(1-\beta)]$, $I_3 = (t(1-\beta), t(1+\beta)]$, $I_4 = (t(1+\beta), 8t]$ and $I_5 = (8t, \infty)$. We have

$$J_1 \equiv t\int_{-\infty}^{0} e^{ty}q(y)dy \leqslant t\int_{-\infty}^{0} e^{ty}dy = 1.$$

On I_5, noting that $c < 1/(8t)$ (provided $\pi < 1/8$), in the light of 7.7.a, we obtain

$$q(y) \leqslant \begin{cases} \exp(-y/(4c)) \leqslant \exp(-2ty), & \text{if } y \geqslant 1/c, \\ \exp\left\{-\frac{y^2}{2}\left(1-\frac{yc}{2}\right)\right\} \leqslant \exp(-y^2/4) \leqslant \exp(-2ty), & \text{if } y < 1/c. \end{cases}$$

Therefore

$$J_5 \equiv t\int_{8t}^{\infty} \mathrm{e}^{ty} q(y)\mathrm{d}y \leqslant t\int_{8t}^{\infty} \mathrm{e}^{-ty}\mathrm{d}y < 1.$$

On I_2 and I_4, we have $y < 8t < 1/c$. Then, by 7.7.a,

$$\begin{aligned} \mathrm{e}^{ty}q(y) &\leqslant \exp\left\{ty - \frac{y^2}{2}(1-\frac{yc}{2})\right\} \\ &\leqslant \exp\left\{ty - \frac{y^2}{2}(1-4tc)\right\} \\ &\equiv \exp\{g(y)\}. \end{aligned}$$

The function $g(y)$ attains its maximum for $y = \frac{t}{1-4tc}$, which, for π (hence c) sufficiently small, lies in I_3. And consequently, for $y \in I_2 \bigcup I_4$,

$$g(y) \leqslant \frac{t^2 - (t - y(1-4tc))^2}{2(1-4tc)} \leqslant \frac{t^2}{2}\left(1 - \frac{\beta^2}{2}\right).$$

Hence

$$J_2 + J_4 \equiv t\Big(\int_0^{t(1-\beta)} + \int_{t(1+\beta)}^{8t}\Big)\mathrm{e}^{ty}q(y)\mathrm{d}y \leqslant 9t^2\exp\left\{\frac{t^2}{2}\left(1-\frac{\beta^2}{2}\right)\right\}.$$

Now let $\alpha = \beta^2/4$. By (29),

$$\begin{aligned} J_2 + J_4 &\leqslant 9t^2\exp\left\{\frac{t^2}{2}(1-\frac{\beta^2}{2})\right\} \\ &\leqslant \frac{9x^2}{(1-\beta)^2}\exp\left\{-\frac{x^2\beta^2}{8(1-\beta)^2}\right\} E\exp\left\{\frac{tS_n}{s_n}\right\}. \end{aligned}$$

Then for $x > x(\beta) = x(\gamma)$ that is large enough, we obtain

$$J_1 + J_5 < 2 < \frac{1}{4}E\mathrm{e}^{tS_n/s_n}, \quad J_2 + J_4 < \frac{1}{4}E\mathrm{e}^{tS_n/s_n}.$$

As a consequence, we have

$$J_3 \equiv t\int_{t(1-\beta)}^{t(1+\beta)} \mathrm{e}^{ty}q(y)\mathrm{d}y > \frac{1}{2}E\mathrm{e}^{tS_n/s_n},$$

which, together with (29), implies that

$$2t^2\beta e^{t^2(1+\beta)}q(x) > \frac{1}{2}\exp\left\{\frac{t^2}{2}(1-\alpha)\right\}.$$

Hence

$$\begin{aligned} q(x) &> \frac{1}{4t^2\beta}\exp\left\{\frac{t^2}{2}\alpha\right\}\exp\left\{-\frac{t^2}{2}(1+2\alpha+2\beta)\right\} \\ &> \exp\left\{-\frac{x^2}{2}\frac{1+2\alpha+2\beta}{(1-\beta)^2}\right\} \end{aligned}$$

provided that $x > x(\gamma)$, and t is large enough. For given $\gamma > 0$, it suffices to choose $\beta > 0$ such that

$$\frac{1+2\beta+\beta^2/2}{(1-\beta)^2} \leqslant 1+\gamma.$$

Therefore, for $\pi(\gamma)$ sufficiently small and $x = x(\gamma)$ sufficiently large,

$$q(x) > \exp\left\{-\frac{x^2}{2}(1+\gamma)\right\}.$$

This completes the proof.

7.7.c (sharpened Kolmogorov upper inequality). In addition to the assumptions in 7.7.a, we assume that for some $\delta \in (0, 1]$,

$$L_n = s_n^{-(2+\delta)}\sum_{i=1}^{n} E|X_i|^{2+\delta} < \infty.$$

Then, for any $x > 0$,

$$P(S_n/s_n \geqslant x) \leqslant \exp\left\{-\frac{x^2}{2} + \frac{1}{6}x^3 c^{1-\delta} L_n e^{xc}\right\}.$$

Proof. For any $t > 0$, we have

$$\begin{aligned} \prod_{j=1}^{n} Ee^{tX_j/s_n} &\leqslant \prod_{j=1}^{n}\left(1+\frac{t^2\sigma_j^2}{2s_n^2}+\left(\sum_{k=3}^{\infty}\frac{t^k c^{k-2-\delta}E|X_j|^{2+\delta}}{k!s_n^{2+\delta}}\right)\right) \\ &\leqslant \exp\left\{\frac{t^2}{2}+\frac{t^3c^{1-\delta}L_n e^{tc}}{6}\right\}, \end{aligned} \tag{30}$$

from which the conclusion follows by taking $t = x$.

7.8 Prokhorov Inequality

Under the conditions of 7.7, for any $x > 0$,

$$P(S_n/s_n \geqslant x) \leqslant \exp\left\{-\frac{x}{2c}\operatorname{arcsinh}\left(\frac{xc}{2}\right)\right\}.$$

Proof. Let $G(x) = \frac{1}{n}\sum_{j=1}^{n} P(X_j < x)$. The d.f. G is concentrated in the interval $[-cs_n, cs_n]$, and

$$\int y^2 \mathrm{d}G(y) = \frac{1}{n}s_n^2.$$

Let G^* be a d.f. with

$$G^*(\{-cs_n\}) = G^*(\{cs_n\}) = 1/(2nc^2), \quad G^*(\{0\}) = 1 - 1/(nc^2).$$

It is easy to verify that

$$\begin{aligned}\int (\cosh ty - 1)\mathrm{d}G(y) &\leqslant \int (\cosh ty - 1)\mathrm{d}G^*(y)\\ &= \frac{1}{nc^2}(\cosh tcs_n - 1).\end{aligned}$$

Then for $t > 0$,

$$\begin{aligned}P(S_n/s_n \geqslant x) &\leqslant \mathrm{e}^{-txs_n} E\mathrm{e}^{tS_n}\\ &= \mathrm{e}^{-txs_n}\prod_{j=1}^{n} E\mathrm{e}^{tX_j}\\ &\leqslant \exp\left\{-txs_n + \sum_{j=1}^{n}(E\mathrm{e}^{tX_j} - 1)\right\}\\ &= \exp\left\{-txs_n + n\int(\mathrm{e}^{ty} - 1 - ty)\mathrm{d}G(y)\right\}\\ &\leqslant \exp\left\{-txs_n + 2n\int(\cosh ty - 1)\mathrm{d}G(y)\right\}\\ &\leqslant \exp\left\{-txs_n + \frac{2}{c^2}(\cosh tcs_n - 1)\right\}.\end{aligned}$$

The exponent achieves its minimum at

$$t = \frac{1}{cs_n}\operatorname{arcsinh}\frac{xc}{2}.$$

We find the minimum value is

$$-\frac{x}{2c}\operatorname{arcsinh}\frac{xc}{2}.$$

7.9 Exponential Inequalities by Censoring

7.9.a (partial sum). Let $X_1, \cdots, X_n$ be i.i.d. r.v.'s, $a > 0$. Define censored r.v.'s

$$Z_j = (-a) \vee (X_j \wedge a).$$

Put $T_n = \sum\limits_{j=1}^{n} Z_j$. Then for any positive numbers r and s,

$$P\left\{|T_n - ET_n| \geqslant \frac{r}{2a}\mathrm{e}^r nE(X_1^2 \wedge a^2) + \frac{sa}{r}\right\} \leqslant 2\mathrm{e}^{-s}.$$

Particularly, for any $\sigma^2 \geqslant E(X_1^2 \wedge a^2)$, $x > 0$,

$$P\left\{|T_n - ET_n| \geqslant \frac{1}{2}\left(1 + \exp\left(\frac{ax}{\sqrt{n}\sigma}\right)\right)x\sqrt{n}\sigma\right\} \leqslant 2\mathrm{e}^{-x^2/2}.$$

Proof. Clearly, it is enough to prove the one-sided inequality

$$P\left\{T_n - ET_n \geqslant \frac{r}{2a}\mathrm{e}^r nE(X_1^2 \wedge a^2) + \frac{sa}{r}\right\} \leqslant \mathrm{e}^{-s}.$$

Put $F(x) = P(X_1 < x)$, $G_+(x) = P(X_1 > x)$, $G_-(x) = P(-X_1 > x)$. Let $t = r/a$. Then

$$\begin{aligned} E\mathrm{e}^{tZ_1} &\leqslant E\left(1 + tZ_1 + \frac{1}{2}t^2 Z_1^2 \mathrm{e}^r\right) \\ &\leqslant \exp\left\{tEZ_1 + \frac{1}{2}\mathrm{e}^r t^2 EZ_1^2\right\}. \end{aligned}$$

The first inequality now follows from

$$\begin{aligned} &P\left\{T_n - ET_n \geqslant \frac{r}{2a}\mathrm{e}^r nE(X_1^2 \wedge a^2) + \frac{sa}{r}\right\} \\ \leqslant &\exp\left\{-tET_n - \frac{tr}{2a}\mathrm{e}^r nE(X_1^2 \wedge a^2) - \frac{tsa}{r} + ntEZ_1 + \frac{n}{2}\mathrm{e}^r t^2 EZ_1^2\right\} \\ = &\mathrm{e}^{-s}. \end{aligned}$$

Letting $s = x^2/2$ and $r = ax/(\sqrt{n}\sigma)$, we obtain the second inequality.

Remark. The conclusions are also true if the censored r.v.'s $Z_j, j = 1, \cdots, n$, are replaced by the truncated r.v.'s

$$Y_j = X_j I(|X_j| \leqslant a), \quad j = 1, \cdots, n,$$

provided $E(X_1^2 \wedge a^2)$ is replaced by $EX_1^2 I(|X_1| \leqslant a)$ in the inequalities.

7.9.b (increments of partial sums, Lin). Let $\{X_n, n \geqslant 1\}$ be a sequence of independent r.v.'s satisfying that there exist positive constants δ, D and σ such that for every j,

$$EX_j = 0, \quad E|X_j|^{2+\delta} \leqslant D, \quad \sigma_{nk}^2 \equiv \sum_{j=n+1}^{n+k} EX_j^2 \geqslant k\sigma^2.$$

Let N and $N_1 = N_1(N)$ be positive integers satisfying that $N_1 \leqslant N$ and $N_1^{2+\delta}/N^2 \to \infty$ as $N \to \infty$. For a given $M > 0$, put

$$Y_j = X_j I(|X_j| < MN^{1/(2+\delta)}) - EX_j I(|X_j| < MN^{1/(2+\delta)}), \quad T_n = \sum_{j=1}^{n} Y_j.$$

Then for any given $0 < \varepsilon < 1$, there exist $C = C(\varepsilon) > 0$, $N_0 = N_0(\varepsilon)$, $x_0 = x_0(\varepsilon)$, such that for any $N \geqslant N_0$, $x_0 \leqslant x \leqslant N_1^{1/2}/N^{1/(2+\delta)}$,

$$P\left\{\max_{1 \leqslant n \leqslant N} \max_{1 \leqslant k \leqslant N_1} |T_{n+k} - T_n|/\sigma_{nN_1} \geqslant (1+\varepsilon)x\right\} \leqslant \frac{CN}{N_1} \mathrm{e}^{-x^2/2}.$$

Proof. Let $\varepsilon > 0$ be given. Define $m_r = [N_1/r]$ and $n_r = [N/m_r + 1]$ for a large positive r to be specified later. Then, by the fact that $EX_j^2 \leqslant \left(E|X_j|^{2+\delta}\right)^{2/(2+\delta)} \leqslant D^{2/(2+\delta)}$, for any $n \in ((j-1)m_r, jm_r]$,

$$\frac{\sigma_{n,N_1}^2}{\sigma_{jm_r,N_1}^2} \leqslant 1 + \frac{m_r D^{2/(2+\delta)}}{N_1 \sigma^2} \leqslant (1+\varepsilon/10)^2,$$

provided r is large enough. Similarly, we can prove the same inequality when $\frac{\sigma_{n,N_1}^2}{\sigma_{jmr,N_1}^2}$ is replaced by $\frac{\sigma_{jmr,N_1}^2}{\sigma_{n,N_1}^2}$ or jm_r is replaced by $(j-1)m_r$. Therefore, we have

$$\begin{aligned} &\max_{1 \leqslant n \leqslant N} \max_{1 \leqslant k \leqslant N_1} |T_{n+k} - T_n|/\sigma_{n,N_1} \\ \leqslant &(1+\varepsilon/10) \max_{1 \leqslant j \leqslant n_r} \max_{1 \leqslant k \leqslant N_1} |T_{jm_r+k} - T_{jm_r}|/\sigma_{jm_r,N_1} \\ &+(1+\varepsilon/10) \max_{1 \leqslant j \leqslant n_r} \max_{1 \leqslant k \leqslant m_r} |T_{jm_r-k} - T_{jm_r}|/\sigma_{(j-1)m_r,N_1}. \end{aligned}$$

Thus,

$$\begin{aligned} &P\{\max_{1 \leqslant n \leqslant N} \max_{1 \leqslant k \leqslant N_1} |T_{n+k} - T_n|/\sigma_{nN_1} \geqslant (1+\varepsilon)x\} \\ \leqslant &P\left\{\max_{1 \leqslant j \leqslant n_r} \max_{1 \leqslant k \leqslant N_1} |T_{jm_r+k} - T_{jm_r}|/\sigma_{jm_r,N_1} \geqslant \left(1+\frac{1}{3}\varepsilon\right)x\right\} \\ &+P\left\{\max_{1 \leqslant j \leqslant n_r} \max_{1 \leqslant k \leqslant m_r} |T_{jm_r-k} - T_{jm_r}|/\sigma_{(j-1)m_r,N_1} \geqslant \frac{1}{3}\varepsilon x\right\}. \quad (31) \end{aligned}$$

By sharpened Kolmogorov upper bound (see 7.7.c) with $c = MN^{1/(2+\delta)}/\sigma_{jm_r,N_1} < M\sigma^{-1}N^{1/(2+\delta)}N_1^{-1/2}$, for $x \leqslant N_1^{1/2}N^{-1/(2+\delta)}$,

$$\begin{aligned}
&P\left\{\max_{1\leqslant j\leqslant n_r}\max_{1\leqslant k\leqslant N_1}|T_{jm_r+k}-T_{jm_r}|/\sigma_{jm_r,N_1}\geqslant\left(1+\frac{1}{3}\varepsilon\right)x\right\}\\
&\leqslant\sum_{j=1}^{n_r}P\left\{\max_{1\leqslant k\leqslant N_1}|T_{jm_r+k}-T_{jm_r}|/\sigma_{jm_r,N_1}\geqslant\left(1+\frac{1}{3}\varepsilon\right)x\right\}\\
&\leqslant n^r\exp\left\{-\frac{x^2\left(1+\frac{1}{3}\varepsilon\right)^2}{2}\left(1-\frac{1}{3}(M/\sigma)^{1-\delta}\mathrm{e}^{M/\delta}D\sigma^{-(2+\delta)}N_1^{-\delta/2}\right)\right\}\\
&\leqslant C_1rNN_1^{-1}\exp\left(-\frac{x^2}{2}\right),
\end{aligned}\tag{32}$$

for all large N so that $(1+\frac{1}{3}\varepsilon)^2(1-\frac{1}{3}(M/\sigma)^{1-\delta}\mathrm{e}^{M/\delta}D\sigma^{-(2+\delta)}N_1^{-\delta/2})\geqslant 1$.

Note that $\sigma_{jm_r,m_r}\leqslant\sqrt{m_r}D^{1/(2+\delta)}$. By taking r so large that

$$\frac{\varepsilon\sigma_{jm_r,N_1}}{3\sigma_{jm_r,m_r}}\geqslant\frac{\varepsilon\sigma\sqrt{N_1}}{3\sqrt{m_r}D^{1/(2+\delta)}}\geqslant\frac{\varepsilon\sigma\sqrt{r}}{3D^{1/(2+\delta)}}\geqslant 1+\frac{1}{3}\varepsilon,$$

and using the same argument as in the proof of (32), we have

$$\begin{aligned}
&P\left\{\max_{1\leqslant j\leqslant n_r}\max_{1\leqslant k\leqslant m_r}|T_{jm_r-k}-T_{jm_r}|/\sigma_{(j-1)m_r,N_1}\geqslant\frac{1}{3}\varepsilon\right\}\\
&\leqslant\sum_{j=1}^{n_r}P\left\{\max_{1\leqslant k\leqslant m_r}|T_{jm_r-k}-T_{jm_r}|/\sigma_{(j-1)m_r,m_r}\geqslant\frac{\varepsilon x\sigma_{(j-1)m_r,N_1}}{3\sigma_{(j-1)m_r,m_r}}\right\}\\
&\leqslant\sum_{j=1}^{n_r}P\left\{\max_{1\leqslant k\leqslant m_r}|T_{jm_r-k}-T_{jm_r}|/\sigma_{(j-1)m_r,m_r}\geqslant\left(1+\frac{1}{3}\varepsilon\right)x\right\}\\
&\leqslant C_2rNN_1^{-1}\exp\left(-\frac{x^2}{2}\right).
\end{aligned}\tag{33}$$

The desired inequality then follows by substituting (32) and (33) into (31).

7.10 Tail Probability of Weighted Sums

Let $\{X_n, n\geqslant 1\}$ be a sequence of i.i.d. r.v.'s with $P(X_1=1)=P(X_1=-1)=1/2$, the so-called Bernoulli (or Rademacher) sequence, $\{\alpha_n, n\geqslant$

1} be a sequence of real numbers with $a \equiv \left(\sum_{n=1}^{\infty} \alpha_n^2\right)^{1/2} < \infty$. Then for any $x > 0$,

$$P\left\{\left|\sum_{n=1}^{\infty} \alpha_n X_n\right| \geqslant x\right\} \leqslant 2\exp(-x^2/2a^2).$$

If there exists a constant $K > 0$ satisfying $x \geqslant Ka$ and $x \max_{n\geqslant 1} |\alpha_n| \leqslant K^{-1}a^2$, then

$$P\left\{\left|\sum_{n=1}^{\infty} \alpha_n X_n\right| \geqslant x\right\} \leqslant \exp(-Kx^2/a^2).$$

Proof. The first inequality is shown by noting that for any $t > 0$,

$$E\exp\left(t\sum_{n=1}^{\infty} \alpha_n X_n\right) \leqslant \prod_{n=1}^{\infty} \exp\left(\frac{t^2}{2}\alpha_n^2\right) = \exp\left(\frac{t^2a^2}{2}\right).$$

And the second one can be deduced from 7.9.a.

Remark. Let $\{Y_n, n \geqslant 1\}$ be a sequence of r.v.'s. We may consider the sum $\sum X_n Y_n$ instead of $\sum Y_n$, where $\{X_n\}$ is a Bernoulli sequence independent of $\{Y_n\}$. The two sums have the same d.f. As the first step, one investigates $\sum X_n Y_n$ given $\{Y_n\}$ conditionally.

References

Bennett G. 1962. Probability inequalities for the sum of independent random variables. Ann. Statist. Assoc., 57: 33-45.

Feller W. 1968. An Introduction to Probability Theory and Its Applications. 3rd ed.. New York: Wiley.

Freedman D A. 1975. On tail probabilities for martingales. Ann. Probab., 3: 100-118.

Hoeffding W. 1963. Probability inequalities for sums of bounded random variables. J. Amer. Statist. Assoc., 58: 13-30.

Lin Z Y. 1987. On Csőrgö-Rèvèsz's increments of sums of non-i.i.d. Random Variables. Scientia Sinica, 30(A): 921-931.

Lin Z Y, Lu C L. 1992. Strong Limit Theorems. Dordrecht & Beijing: Kluwer Academic Publishers & Science Press.

Loéve M. 1977. Probability Theory. 4th ed.. New York: Springer-Verlag.

Petrov V V. 1995. Limit Theorems of Probability Theory: Sequences of Independent Random Variables. Oxford & New York: Clarendon Press & Oxford University Press.

Shorack G R, Wellner J A. 1986. Empirical processes with applications to statistics. New York: Wiley.

Stout W. 1974. Almost Sure Convergence. New York: Academic Press.

Chapter 8

Moment Inequalities Related to One or Two Variables

In this chapter, we shall introduce some important moment inequalities, such as Hölder, Cauchy-Schwarz and Jensen inequalities. These inequalities not only frequently used in research works, but also help to embedding random variables into some linear spaces so that many probabilistic problems can be understood and analyzed geometrically.

The following moment inequalities are given by the expectation $E(\cdot)$, but they are also true if we replace $E(\cdot)$ by the conditional expectation $E(\cdot|\mathscr{A})$.

8.1 Moments of Truncation

For any positive numbers r and C,

$$E|XI(|X| \leqslant C)|^r \leqslant E|X|^r.$$

Proof.

$$\begin{aligned} E|X|^r &= \int P\{|X^r > x\}\,\mathrm{d}x \\ &\geqslant \int P\{|X|^r I(|X| \leqslant C) > x\}\,\mathrm{d}x = E|XI(|X| \leqslant C)|^r. \end{aligned}$$

8.2 Exponential Moment of Bounded Variables

If $X \leqslant 1$ a.s., then

$$E(\exp X) \leqslant \exp(EX + EX^2).$$

Proof. The inequality follows from the elementary inequality

$$\mathrm{e}^x \leqslant 1 + x + x^2 \quad \text{for all } x \leqslant 1.$$

8.3 Hölder Type Inequalities

8.3.a (Hölder inequality). For $p > 1$ and $q > 1$ with $\frac{1}{p} + \frac{1}{q} = 1$,

$$E|XY| \leqslant (E|X|^p)^{\frac{1}{p}}(E|Y|^q)^{\frac{1}{q}}.$$

Proof. We may assume that $0 < E|X|^p,\ E|Y|^q < \infty$, since the inequality is trivial otherwise. Note that $-\log x$ is a convex function on $(0, \infty)$. Hence, for $a, b > 0$ we have

$$-\log\left(\frac{a^p}{p} + \frac{b^q}{q}\right) \leqslant -\frac{1}{p}\log a^p - \frac{1}{q}\log b^q = -\log ab,$$

or equivalently

$$ab \leqslant \frac{a^p}{p} + \frac{b^q}{q}, \quad 0 \leqslant a, b \leqslant \infty.$$

Thus

$$\begin{aligned} &E(|X|/(E|X|^p)^{1/p})(|Y|/(E|Y|^q)^{1/q}) \\ &\leqslant \frac{1}{p}E(|X|/(E|X|^p)^{1/p})^p + \frac{1}{q}E(|Y|/(E|Y|^q)^{1/q})^q \\ &= \frac{1}{p} + \frac{1}{q} = 1, \end{aligned}$$

which implies the desired inequality.

8.3.b (Cauchy-Schwarz inequality).

$$E|XY| \leqslant (EX^2)^{1/2}(EY^2)^{1/2}.$$

Proof. Let $p = q = 2$ in the Hölder inequality.

8.3.c (Lyapounov inequality). Put $\beta_p = E|X|^p$. For $0 \leqslant r \leqslant s \leqslant t$,

$$\beta_s^{t-r} \leqslant \beta_r^{t-s}\beta_t^{s-r}.$$

Proof. Let $p = \frac{t-r}{t-s}$, $q = \frac{t-r}{s-r}$ and let X and Y in 8.3.a be replaced by $|X|^{\frac{t-s}{t-r}r}$ and $|X|^{\frac{s-r}{t-r}t}$ respectively.

Remark. From Lyapounov's inequality, we have

$$\log \beta_s \leqslant \frac{t-s}{t-r}\log \beta_r + \frac{s-r}{t-r}\log \beta_t.$$

That is to say, $\log \beta_s$ is a convex function in s.

8.3.d (a generalization of Hölder's inequality). For $0 < p < 1$ and $q = -p/(1-p)$,

$$E|XY| \geqslant (E|X|^p)^{1/p}(E|Y|^q)^{1/q}.$$

Proof. Put $X' = |XY|^p$, $Y' = |Y|^{-p}$. Then by the Hölder inequality

$$E|X|^p = EX'Y' \leqslant (EX'^{1/p})^p(EY'^{1/(1-p)})^{1-p} = (E|XY|)^p(E|Y|^q)^{1-p},$$

which implies the desired inequality.

8.4 Jensen Type Inequalities

8.4.a (Jensen inequality). Let g be a convex function on $\mathbb{R}$. Suppose that expectations of X and $g(X)$ exist. Then

$$g(EX) \leqslant Eg(X).$$

Equality holds for strictly convex g if and only if $X = EX$ a.s.

Proof. We need the following property of a convex function on $\mathbb{R}$: for any x and y,

$$g(x) \geqslant g(y) + (x-y)g_r'(y), \tag{34}$$

where g_r' is the right derivative of g (c.f., e.g., Hardy et. al. (1934, 91-95)). Let $x = X$, $y = EX$ in the above inequality. We obtain

$$g(X) \geqslant g(EX) + (X - EX)g_r'(EX).$$

Taking expectation yields the desired inequality.

When g is strictly convex, g_r' is strictly increasing. If $X = EX$ a.s., $Eg(X) = g(EX)$, obviously. On the contrary, if $Eg(X) = g(EX)$, by letting $x = EX$, $y = X$ in (34), we obtain

$$E(EX - X)g_r'(X) = 0 \quad \Rightarrow \quad E(EX - X)(g_r'(X) - g_r'(E(X))) = 0,$$

which implies that $X = EX$ a.s.

8.4.b (monotonicity of L_r-norm——consequence of the Jensen inequality). For any $0 < r \leqslant s$,

$$(E|X|^r)^{1/r} \leqslant (E|X|^s)^{1/s}.$$

In particular, for any $p \geqslant 1$,

$$E|X| \leqslant (E|X|^p)^{1/p}.$$

Proof. Take $g(x) = |x|^{s/r}$ and replace X by $|X|^r$ in the Jensen inequality.

Remark. This inequality shows that $\beta_s^{1/s}$ is an increasing function in s.

8.4.c (arithmetic-geometric inequality). Let X be a non-negative r.v., then

$$EX \geqslant \exp\{E \log X\}.$$

Unless X is a degenerate r.v. or $E \log X = \infty$, the inequality is strict.

Proof. Apparently we only need to consider the case where $EX < \infty$. The inequality is a direct consequence of 8.4.a, setting g as: when $x > 0$, $g(x) = -\log x$; when $x \leqslant 0$, $g(x) = \infty$.

Remark. A formal generalization of the inequality above is: let X be a non-negative r.v., then

$$(EX^r)^{1/r} \geqslant \exp\{E \log x\}, \quad r \geqslant 1;$$
$$(EX^r)^{1/r} \leqslant \exp\{E \log x\}, \quad 0 < r < 1.$$

8.5 Dispersion Inequality of Censored Variables

Put $Y = aI(X < a) + XI(a \leqslant X \leqslant b) + bI(X > b)$ $(-\infty \leqslant a < b \leqslant \infty)$. Then, if $E|X| < \infty$, for $1 \leqslant p < \infty$,

$$E|X - EX|^p \geqslant E|Y - EY|^p.$$

Proof. Put $\alpha(t) = a \vee (t \wedge b)$ and $\beta(t) = t - \alpha(t) - EX + E\alpha(X)$, where both are nondecreasing in t. By inequality (34) for a convex function used in 8.4.a with $g(x) = |x|^p$, we have

$$|X - EX|^p \geqslant |\alpha(X) - E\alpha(X)|^p + \beta(X) g_r^{'}(\alpha(X) - E\alpha(X)).$$

Since both $g_r^{'}(\alpha(t) - E\alpha(X))$,and $\beta(t)$ are nondecreasing,

$$E\beta(X) g_r^{'}(\alpha(X) - E\alpha(X)) \geqslant E\beta(X) E g_r^{'}(\alpha(X) - E\alpha(X)) = 0.$$

Taking expectation in the above two inequalities yields the desired conclusion.

Remark. As a consequence, we have Var $X^+ \leqslant$ Var X, Var $X^- \leqslant$ Var X.

8.6 Monotonicity of Moments of Sums

If X and Y are two r.v.'s with $E|X|^r < \infty$, $E|Y|^r < \infty$, $r \geqslant 1$, and $E(Y|X) = 0$ a.s., then

$$E|X+Y|^r I_A \geqslant E|X|^r I_A,$$

where A is a event defined on X. In particular, if, in addition, X and Y are independent with $EY = 0$, then

$$E|X+Y|^r \geqslant E|X|^r. \tag{35}$$

Proof. The desired inequality follows from the following (conditional) Jensen inequality,

$$|X|^r I_A = |E((X+Y)I_A|X)|^r \leqslant E(|X+Y|^r I_A|X).$$

Remark. In general, if $f(x)$ is a convex function on $[0,\infty)$, then

$$Ef(|X+Y|I_A) \geqslant Ef(|X|I_A).$$

This follows from the conditional Jenssen inequality $f(|X|I_A) = f(|E((X+Y)I_A|X)|) \leqslant E[f(|X+Y|I_A)|X]$.

8.7 Symmetrization Moment Inequalities

Let X and X' be iid. r.v.'s satisfying $E|X|^r < \infty$ for some r.

8.7.a. $\frac{1}{2}E|X-X'|^r \leqslant E|X|^r$ for $0 < r \leqslant 2$.

Proof. For the case of $r = 2$ the inequality is trivial. If $0 < r < 2$, denote the d.f. and c.f. of X by $F(x)$ and $f(t)$ respectively. We have the formula

$$|x|^r = K(r)\int (1-\cos xt)/|t|^{r+1}\mathrm{d}t, \quad 0<r<2,$$

where x is a real number and

$$K(r) = \left(\int \frac{1-\cos u}{|u|^{r+1}}\mathrm{d}u\right)^{-1} = \frac{\Gamma(r+1)}{\pi}\sin\frac{r\pi}{2}.$$

Then

$$\begin{aligned} E|X|^r &= \int |x|^r \mathrm{d}F(x) = K(r)\int\int \frac{1-\cos xt}{|t|^{r+1}}\mathrm{d}t\mathrm{d}F(x) \\ &= K(r)\int \frac{1-\mathrm{Re}f(t)}{|t|^{r+1}}\mathrm{d}t. \end{aligned} \tag{36}$$

Using this and the identity $2(1-\mathrm{Re}f(t)) = (1-|f(t)|^2)+|1-f(t)|^2$, we obtain

$$E|X|^r = \frac{1}{2}E|X-X'|^r + \frac{1}{2}K(r)\int \frac{|1-f(t)|^2}{|t|^{r+1}}\mathrm{d}t, \quad 0<r<2,$$

which implies the desired inequality.

8.7.b. For any a,

$$\frac{1}{2}E|X-mX|^r \leqslant E|X-X'|^r \leqslant 2c_r E|X-a|^r \quad \text{for } r>0.$$

Proof. By the elementary inequality $|\alpha+\beta|^r \leqslant c_r(|\alpha|^r + |\beta|^r)$, where $c_r = 1$ or 2^{r-1} in accordance with $r \leqslant 1$ or not, we have

$$\begin{aligned} E|X-X'|^r &= E|(X-a)-(X'-a)|^r \leqslant c_r(E|X-a|^r + E|X'-a|^r) \\ &= 2c_r E|X-a|^r. \end{aligned}$$

This is the RHS. As for the LHS, it is trivial if $E|X-X'|^r = \infty$. Then, according to the inequality just proven (with $a = mX$), $E|X-mX|^r = \infty$. Thus we can assume that $E|X-X'|^r < \infty$. Let

$$q(x) = P\{|X-mX| \geqslant x\} \quad \text{and} \quad q^s(x) = P\{|X-X'| \geqslant x\}.$$

By the weak symmetrization inequality 5.3.a, $q(x) \leqslant 2q^s(x)$. Therefore it follows, upon integration by parts, that

$$E|X-mX|^r = \int_0^\infty rx^{r-1}q(x)\mathrm{d}x \leqslant 2\int_0^\infty rx^{r-1}q^s(x)\mathrm{d}x = 2E|X-X'|^r$$

as desired.

8.8 Kimball Inequality

Let $u(x)$ and $v(x)$ be both nonincreasing or both nondecreasing functions. Then

$$Eu(X)Ev(X) \leqslant E(u(X)v(X)),$$

if the indicated expectations exist.

Proof. We use the Hoeffding lemma.

$$\begin{aligned} \mathrm{Cov}(u(X),v(X)) = \int\int \{&P(u(X)<s, v(X)<t) \\ &-P(u(X)<s)P(v(X)<t)\}\mathrm{d}s\mathrm{d}t. \qquad (37) \end{aligned}$$

Since $u(x)$ and $v(x)$ are both nonincreasing or both nondecreasing, we have

$$\begin{aligned} P(u(X) < s, v(X) < t) &= \min\{P(u(X) < s), P(v(X) < t)\} \\ &\geqslant P(u(X) < s)P(v(X) < t). \end{aligned}$$

Inserting it into (37) yields the desired inequality.

8.9 Exponential Moment of Normal Variable

Let $1 \leqslant p < 2$, X be a standard normal r.v. Then,

$$\begin{aligned} &\exp(t^{\frac{2}{2-p}}\beta_p) \leqslant E\exp(t|X|^p) \\ &\leqslant \exp\left\{t\delta_p + t^{\frac{2}{2-p}}\beta_p + t^{\frac{3}{2}}\frac{9}{(2-p)^2}\right\} \\ &\wedge \exp\{(1+\varepsilon)t\delta_p + t^{\frac{2}{2-p}}(\beta_p + c(\varepsilon,p))\} \end{aligned}$$

for each $t > 0, \varepsilon > 0$ where

$$\delta_p = E|X|^p, \quad \beta_p = p^{\frac{p}{2-p}}\frac{2-p}{2}, \quad c(\varepsilon,p) = \left(\frac{18}{2-p}\right)^{\frac{8}{2-p}}\left(\frac{1}{\varepsilon}\right)^{\frac{3p-2}{2-p}}.$$

Proof. Let $f(x) = \exp(-\frac{1}{2}x^2 + tx^p)$. It is clear that $f(x)$ is increasing on $(0, (pt)^{1/(2-p)})$ and decreasing on $((pt)^{1/(2-p)}, \infty)$. Hence we have

$$\begin{aligned} f(x) &\leqslant \exp\{t(pt)^{p/(2-p)} - \frac{1}{2}(pt)^{2/(2-p)}\} \\ &= \exp(t^{\frac{2}{2-p}}\beta_p) \quad \text{for } 0 \leqslant x \leqslant (pt)^{1/(2-p)}, \end{aligned}$$

$$\begin{aligned} &f(x + (pt)^{\frac{1}{2-p}}) \\ &= \exp\left\{t(pt)^{\frac{p}{2-p}}\left(1 + \frac{x}{(pt)^{1/(2-p)}}\right)^p - \frac{1}{2}(x + (pt)^{\frac{1}{2-p}})^2\right\} \\ &\leqslant \exp\left\{t(pt)^{\frac{p}{2-p}}\left(1 + \frac{px}{(pt)^{1/(2-p)}} + \frac{p(p-1)x^2}{2(pt)^{2/(2-p)}}\right) - \frac{1}{2}(x + (pt)^{\frac{1}{2-p}})^2\right\} \\ &= \exp(t^{\frac{2}{2-p}}\beta_p)\exp\left(-\frac{2-p}{2}x^2\right) \quad \text{for } x > 0. \end{aligned}$$

Moreover we have

$$e^x \leqslant 1 + x + \frac{1}{2}x^{3/2}e^x \quad \text{for } x \geqslant 0.$$

If $pt \leqslant 1$, by the above three inequalities we have

$$
\begin{aligned}
& E\exp(t|X|^p) \leqslant 1 + tE|X|^p + \frac{t^{3/2}}{2}E|X|^{3p/2}\mathrm{e}^{t|X|^p} \\
&= 1 + t\delta_p + \sqrt{\frac{2}{\pi}}t^{3/2}\left\{\int_0^{(pt)^{1/(2-p)}} x^{3p/2}f(x)\mathrm{d}x\right. \\
&\quad \left. + \int_0^\infty (x + (pt)^{1/(2-p)})^{3p/2} f(x + (pt)^{1/(2-p)})\mathrm{d}x\right\} \\
&= 1 + t\delta_p + \sqrt{\frac{2}{\pi}}t^{3/2}\exp(t^{2/(2-p)}\beta_p) \\
&\quad \cdot \left\{1 + 2^{3p/2-1}\int_0^\infty (1 + x^{3p/2})\exp\left(-\frac{2-p}{2}x^2\right)\mathrm{d}x\right\} \\
&\leqslant 1 + t\delta_p + t^{3/2}\exp(t^{2/(2-p)}\beta_p)\left(1 + \frac{2}{(2-p)^{1/2}} + \frac{6}{(2-p)^{3p/4+1/2}}\right) \\
&\leqslant 1 + t\delta_p + \frac{9t^{3/2}}{(2-p)^2}\exp(t^{2/(2-p)}\beta_p) \\
&\leqslant \exp\left\{t\delta_p + t^{2/(2-p)}\beta_p + \frac{9t^{3/2}}{(2-p)^2}\right\}.
\end{aligned}
$$

Similarly, for $pt > 1$, we have

$$
\begin{aligned}
& E\exp(t|X|^p) \\
&= \frac{2}{\sqrt{2\pi}}\left\{\int_0^{(pt)^{1/(2-p)}} f(x)\mathrm{d}x + \int_0^\infty f(x + (pt)^{1/(2-p)})\mathrm{d}x\right\} \\
&\leqslant \exp(t^{2/(2-p)}\beta_p)\left\{(pt)^{1/(2-p)} + \frac{2}{\sqrt{2\pi}}\int_0^\infty \exp\left(-\frac{2-p}{2}x^2\right)\mathrm{d}x\right\} \\
&= \exp(t^{2/(2-p)}\beta_p)\{(pt)^{1/(2-p)} + (2-p)^{-1/2}\} \\
&\leqslant \exp\left\{t^{2/(2-p)}\beta_p + \frac{2}{3(2-p)}\log(pt)^{3/2} + \frac{1}{2}\log\frac{1}{2-p}\right\} \\
&\leqslant \exp\left\{t^{2/(2-p)}\beta_p + \frac{2(pt)^{3/2}}{3(2-p)} + \frac{(pt)^{3/2}}{2(2-p)}\right\} \\
&\leqslant \exp\left\{t^{2/(2-p)}\beta_p + \frac{8}{2-p}t^{3/2}\right\} \\
&\leqslant \exp\left\{t^{2/(2-p)}\beta_p + t\delta_p + \frac{9}{(2-p)^2}t^{3/2}\right\}.
\end{aligned}
$$

By these two inequalities we obtain

$$E\exp(t|X|^p) \leqslant \exp\left\{t\delta_p + t^{2/(2-p)}\beta_p + t^{3/2}\frac{9}{(2-p)^2}\right\}.$$

Now we show another upper bound. Noting that $\delta_p \geqslant (E|X|)^p = (\sqrt{2/\pi})^p \geqslant 1/2$, we have

$$\begin{aligned}\frac{9}{(2-p)^2}t^{3/2} &\leqslant \frac{\varepsilon t}{2} + \frac{9}{(2-p)^2}\left(\frac{18}{\varepsilon(2-p)^2}\right)^{\frac{4}{2-p}-3} t^{\frac{2}{2-p}} \\ &\leqslant \varepsilon\delta_p t + \left(\frac{18}{2-p}\right)^{\frac{8}{2-p}}\left(\frac{1}{\varepsilon}\right)^{\frac{3p-2}{2-p}}\beta_p t^{\frac{2}{2-p}},\end{aligned}$$

which, in combination with the first upper bound, implies the second one.

Consider the lower bound. We have

$$\begin{aligned}&E\exp(t|X|^p) \\ &= \frac{2}{\sqrt{2\pi}}\int_0^{(pt)^{\frac{1}{2-p}}} \exp\left(tx^p - \frac{x^2}{2}\right)\mathrm{d}x + \frac{2}{\sqrt{2\pi}}\int_{(pt)^{\frac{1}{2-p}}}^{\infty} \exp\left(tx^p - \frac{x^2}{2}\right)\mathrm{d}x \\ &\geqslant \frac{2}{\sqrt{2\pi}}\int_0^{(pt)^{\frac{1}{2-p}}} \exp\left\{t(pt)^{\frac{p}{2-p}}\left(1 - \frac{x}{(pt)^{\frac{1}{2-p}}}\right)^p - \frac{1}{2}\left((pt)^{\frac{1}{2-p}} - x\right)^2\right\}\mathrm{d}x \\ &\quad + \frac{2}{\sqrt{2\pi}}\exp\left(t(pt)^{\frac{p}{2-p}}\right)\int_{(pt)^{\frac{1}{2-p}}}^{\infty}\exp\left(-\frac{1}{2}x^2\right)\mathrm{d}x.\end{aligned}$$

Hence, noting that for $0 \leqslant x \leqslant (pt)^{1/(2-p)}$,

$$\begin{aligned}&t(pt)^{\frac{p}{2-p}}\left(1 - \frac{x}{(pt)^{1/(2-p)}}\right)^p - \frac{1}{2}((pt)^{1/(2-p)} - x)^2 \\ &\geqslant t(pt)^{\frac{p}{2-p}}\left(1 - \frac{px}{(pt)^{1/(2-p)}}\right) - \frac{1}{2}((pt)^{1/(2-p)} - x)^2 \\ &= t^{\frac{2}{2-p}}\beta_p - \frac{1}{2}x^2.\end{aligned}$$

We conclude that

$$\begin{aligned} E\exp(t|X|^p) \geqslant & \frac{2}{\sqrt{2\pi}} \exp(t^{\frac{2}{2-p}}\beta_p) \int_0^{(pt)^{1/(2-p)}} \exp\left(-\frac{x^2}{2}\right) dx \\ & + \frac{2}{\sqrt{2\pi}} \exp(t^{\frac{2}{2-p}} p^{\frac{p}{2-p}}) \int_{(pt)^{\frac{1}{2-p}}}^{\infty} \exp\left(-\frac{x^2}{2}\right) dx \\ \geqslant & \exp(t^{\frac{2}{2-p}}\beta_p), \end{aligned}$$

as desired.

See Csáki, Csörgő and Shao (1995).

8.10 Inequalities of Nonnegative Variable

Let X be a nonnegative r.v.

8.10.a. Let $d > 0$ be an integer. The following are equivalent:

(i) There is a $K > 0$ such that for any $p \geqslant 2$,

$$(EX^p)^{1/p} \leqslant Kp^{d/2}(EX^2)^{1/2}.$$

(ii) For some $t > 0$,

$$E\exp(tX^{2/d}) < \infty.$$

Proof. The equivalence is obtained via the Taylor expansion of the exponential function in (ii).

8.10.b. Let a and b be positive numbers. The following are equivalent:

(1) There is a $K > 0$ such that for any $x > 0$,

$$P\{X > K(b + ax)\} \leqslant K\exp(-x^2/K).$$

(2) There is a $K > 0$ such that for any $p \geqslant 1$,

$$(EX^p)^{1/p} \leqslant K(b + a\sqrt{p}).$$

Proof. (1)$\Leftrightarrow$(ii) with $d = 1$ $\Leftrightarrow$(i) with $d = 1$ $\Leftrightarrow$(2).

8.10.c (Tong). If $EX^k < \infty$, then

$$EX^k \geqslant (EX^{k/r})^r \geqslant (EX)^k + (EX^{k/r} - (EX)^{k/r})^r \tag{38}$$

holds for all $k > r \geqslant 2$; a special form is

$$EX^k \geqslant (EX)^k + (\mathrm{Var}X)^{k/2}. \tag{39}$$

Equality holds if and only if $X = EX$ a.s.

Proof. The first inequality in (38) follows from Lyapounov's inequality 8.3.c while the second in (38) follows from the following elementary inequality: for any $a > b > 0$,

$$(a+b)^r - a^r = rb\xi^{r-1} \geqslant b^r \qquad \text{for any } r \geqslant 1, \tag{40}$$

where $\xi \geqslant a > b$.

The inequality (39) immediately follows from (38) with $r = k/2$. By 8.4.a, the equality signs hold if and only if $X = EX$ a.s.

8.11 Freedman Inequality

Let X be a r.v. with $|X| \leqslant 1$, $EX = 0$ and $\sigma^2 = EX^2$, λ be a positive number. Then

$$\exp\{(\mathrm{e}^{-\lambda} - 1 + \lambda)\sigma^2\} \leqslant E\exp(\lambda X) \leqslant \exp\{(\mathrm{e}^{\lambda} - 1 - \lambda)\sigma^2\}.$$

Proof. We first prove the RHS inequality. By Taylor's expansion,

$$\begin{aligned} E\exp(\lambda X) &= 1 + \frac{1}{2}\lambda^2\sigma^2 + \sum_{k=3}^{\infty}\frac{\lambda^k}{k!}EX^k \leqslant 1 + \frac{1}{2}\lambda^2\sigma^2 + \sum_{k=3}^{\infty}\frac{\lambda^k}{k!}\sigma^2 \\ &\leqslant \exp\left(\sum_{k=2}^{\infty}\frac{\lambda^k}{k!}\sigma^2\right) = \exp\left(\sigma^2(\mathrm{e}^{\lambda} - 1 - \lambda)\right). \end{aligned}$$

Next, we show the LHS inequality. To this end, set

$$g(\lambda) = E\mathrm{e}^{\lambda X} - \exp\left(\sigma^2(\mathrm{e}^{-\lambda} - 1 + \lambda)\right).$$

It is easy to verify that

$$\begin{aligned} g'(\lambda) &= EX\mathrm{e}^{\lambda X} - \sigma^2(1 - \mathrm{e}^{-\lambda})\exp\left(\sigma^2(\mathrm{e}^{-\lambda} - 1 + \lambda)\right), \\ g''(\lambda) &= EX^2\mathrm{e}^{\lambda X} - \left(\sigma^2\mathrm{e}^{-\lambda} - \sigma^4(1 - \mathrm{e}^{-\lambda})^2\right)\exp\left(\sigma^2(\mathrm{e}^{-\lambda} - 1 + \lambda)\right). \end{aligned}$$

Note that $g(0) = g'(0) = g''(0) = 0$. To show $g(\lambda) \geqslant 0$ for all $\lambda > 0$, we only need to show that $g''(\lambda) \geqslant 0$ for all $\lambda > 0$.

Noticing $X \geqslant -1$ and thus $EX^2 e^{\lambda X} - \sigma^2 e^{-\lambda} = EX^2(e^{\lambda X} - e^{-\lambda}) \geqslant 0$ for all $\lambda > 0$, we only need to show that

$$\sigma^2 e^{-\lambda} - \left(\sigma^2 e^{-\lambda} - \sigma^4(1-e^{-\lambda})^2\right)\exp\left(\sigma^2(e^{-\lambda} - 1 + \lambda)\right) \geqslant 0.$$

This is equivalent to

$$h(\lambda) \equiv \exp\left(-\sigma^2(e^{-\lambda} - 1 + \lambda)\right) - \left(1 - \sigma^2(1-e^{-\lambda})^2 e^{\lambda}\right) \geqslant 0. \quad (41)$$

It is easy to verify that

$$\begin{aligned} h'(\lambda) &= \sigma^2(e^{-\lambda} - 1)\exp\left(-\sigma^2(e^{-\lambda} - 1 + \lambda)\right) + \sigma^2(e^{\lambda} - e^{-\lambda}) \\ &\geqslant \sigma^2(1-e^{-\lambda})\left(1 - \exp\left(-\sigma^2(e^{-\lambda} - 1 + \lambda)\right)\right) \geqslant 0, \end{aligned}$$

where the last step follows from the fact that $e^{-\lambda} - 1 + \lambda > 0$ for all $\lambda > 0$. (41) follows from the above and $h(0) = 0$. Consequently, the proof of the LHS inequality is complete.

8.12 Exponential Moment of Upper Truncated Variable

Let X be a r.v. with $EX = 0$. Let $a > 0$ and $0 \leqslant \alpha \leqslant 1$. Then for any $t \geqslant 0$, we have

$$E\exp\{tXI(X \leqslant a)\} \leqslant \exp\left\{\frac{t^2}{2}EX^2 + \frac{t^{2+\alpha}e^{ta}E|X|^{2+\alpha}}{(\alpha+1)(\alpha+2)}\right\}.$$

Proof. For $u \leqslant u_0$, we have

$$\begin{aligned} e^u - 1 - u - \frac{1}{2}u^2 &= \int_0^u \int_0^s (e^w - 1)dwds \leqslant \left|\int_0^u \int_0^s |w|^\alpha e^{u_0} dwds\right| \\ &= \frac{|u|^{2+\alpha}e^{u_0}}{(\alpha+1)(\alpha+2)}. \end{aligned}$$

Thus

$$\begin{aligned} &E\exp\{tXI(X \leqslant a)\} \\ &= 1 + tEXI(X \leqslant a) + \frac{t^2}{2}EX^2 I(X \leqslant a) + \frac{t^{2+\alpha}e^{ta}E|X|^{2+\alpha}}{(\alpha+1)(\alpha+2)} \\ &\leqslant 1 + \frac{t^2}{2}EX^2 + \frac{t^{2+\alpha}e^{ta}E|X|^{2+\alpha}}{(\alpha+1)(\alpha+2)} \\ &\leqslant \exp\left\{\frac{t^2}{2}EX^2 + \frac{t^{2+\alpha}e^{ta}E|X|^{2+\alpha}}{(\alpha+1)(\alpha+2)}\right\}, \end{aligned}$$

as desired.

References

Chow Y S, Teicher H. 1997. Probability Theory: Independence, Interchangeability, Martingales. 2nd ed., New York: Springer-Verlag.

Csáki E, Csörgő M, Shao Q M. 1995. Moduli of continuity for l^p-valued Gaussian processes. Acta Sci. Math., 60: 149-175.

Feller W. 1968. An Introduction to Probability Theory and Its Applications. 3rd ed.. New York: Wiley.

Hardy G H, Littlewood J E, Polya G. 1952. Inequalities. Cambridge: Cambridge University Press.

Kimball A W. 1951. On dependent tests of significance in the analysis of variance. Ann. Math. Statist., 22: 600-602.

Petrov V V. 1995. Limit Theorems of Probability Theory: Sequences of Independent Random Variables. Oxford & New York: Clarendon Press & Oxford University Press.

Shorack G R, Wellner J A. 1986. Empirical processes with applications to statistics. New York: Wiley.

Chapter 9

Moment Estimates of (Maximum of) Sums of Random Variables

Limiting properties of partial sums of a sequence of random variables form one of the largest subjects of probability theory. Therefore, moment estimation of the (maximal) sum of random variables is very important in the research of limiting theorems. In this chapter, we introduce some most important inequalities, such as von Bahr-Esseen, Khintchine, Marcinkiewics-Zygmund-Berkholder inequalities. Proofs of such inequalities are rather involved. Some simpler proofs will be provided in this Chapter. For those with complicated proofs, the references will be given therein.

9.1 Elementary Inequalities

Let $X_1, \cdots, X_n$ be r.v.'s and $S_n = \sum\limits_{j=1}^{n} X_j$.

9.1.a. c_r-inequality

$$E\left|S_n\right|^r \leqslant c_r \sum_{i=1}^{n} E|X_i|^r,$$

where $c_r = 1$ or n^{r-1} according whether $0 < r \leqslant 1$ or $r > 1$.

Proof. We use the method of probability to prove the desired inequality. First put $r > 1$. Let ξ be a r.v. which has values $a_1, \cdots, a_n$ in equal probability, then

$$E|\xi| = \frac{1}{n}\sum_{i=1}^{n}|a_i|, \quad E|\xi|^r = \frac{1}{n}\sum_{i=1}^{n}|a_i|^r.$$

Using the Jensen inequality, we can easily obtain

$$\frac{1}{n^r}\left(\sum_{i=1}^{n}|a_i|\right)^r \leqslant \frac{1}{n}\sum_{i=1}^{n}|a_i|^r.$$

When $0 < r \leqslant 1$, it is only necessary to prove for the situation where $a_1, \cdots, a_n$ not all equals 0. Here

$$|a_k| \Big/ \sum_{i=1}^{n}|a_i| \leqslant |a_k|^r \Big/ \left(\sum_{i=1}^{n}|a_i|\right)^r, \qquad k = 1, \cdots, n.$$

By adding all the inequalities above we can obtain the desired inequality.
9.1.b. Let $1 \leqslant r \leqslant 2$. If $X_1, \cdots, X_n$ are independent and symmetric r.v.'s (or, more generally, the conditional d.f. of X_{j+1} given S_j is symmetric, $1 \leqslant j \leqslant n-1$), then

$$E|S_n|^r \leqslant \sum_{j=1}^{n} E|X_j|^r. \tag{42}$$

Proof. The inequality is trivial when $n = 1$. We shall prove the inequality by induction. Fix m, $1 \leqslant m < n$ and let $f_m(t)$ be the c.f. of X_{m+1} (in the general case, $f_m(t) = E(\exp(itX_{m+1})|S_m)$). According to the symmetry assumption, $f_m(t)$ is real. The conditional c.f. of S_{m+1} given S_m is $\exp(itS_m)f_m(t)$. Applying (36), we have

$$E(|S_{m+1}|^r|S_m) = K(r)\int \frac{1-\cos(tS_m)f_m(t)}{|t|^{r+1}}\mathrm{d}t \quad \text{a.s.}$$

But

$$\begin{aligned}
&1-\cos(tS_m)f_m(t)\\
=&(1-\cos(tS_m)) + (1-f_m(t)) - (1-\cos(tS_m))(1-f_m(t))\\
\leqslant&(1-\cos(tS_m)) + (1-f_m(t)).
\end{aligned}$$

Therefore

$$\begin{aligned}
E(|S_{m+1}|^r|S_m) \leqslant& K(r)\int \frac{1-\cos(tS_m)}{|t|^{r+1}}\mathrm{d}t + K(r)\int \frac{1-f_m(t)}{|t|^{r+1}}\mathrm{d}t\\
=&|S_m|^r + E(|X_{m+1}|^r|S_m).
\end{aligned}$$

Taking expectation, we obtain $E|S_{m+1}|^r \leqslant E|S_m|^r + E(|X_{m+1}|^r)$. The inequality is thus proved by induction.

9.2 Minkowski Type Inequalities

9.2.a (Minkowski inequality).

$$\left(E\left|\sum_{j=1}^{n} X_j\right|^r\right)^{1/r} \leqslant \sum_{j=1}^{n}(E|X_j|^r)^{1/r} \quad \text{for } r \geqslant 1;$$

$$\left(E\left(\sum_{j=1}^{n} |X_j|\right)^r\right)^{1/r} > \sum_{j=1}^{n}(E|X_j|^r)^{1/r} \quad \text{for } 0 < r < 1.$$

Proof. Clearly, it is enough to consider the case where $n = 2$. Let $r > 1$. By the Hölder inequality 8.3.a,

$$\begin{aligned} E|X_1 + X_2|^r &\leqslant E(|X_1||X_1 + X_2|^{r-1}) + E(|X_2||X_1 + X_2|^{r-1}) \\ &\leqslant ((E|X_1|^r)^{1/r} + (E|X_2|^r)^{1/r})(E|X_1 + X_2|^r)^{(r-1)/r}. \end{aligned}$$

Dividing both sides by $(E|X_1 + X_2|^r)^{(r-1)/r}$ yields the first inequality.

Similarly we can prove the second one by using 8.3.d instead of 8.3.a.

Remark. If fact, when $r \geqslant 1$, $(E|X|^r)^{1/r}$ can be regarded as a norm of X. Therefore, the first inequality is just a consequence of the triangular inequality.

9.2.b (companion to the Minkowski inequality).

$$E\left(\sum_{j=1}^{n} |X_j|\right)^r \geqslant \sum_{j=1}^{n} E|X_j|^r \quad \text{for } r \geqslant 1;$$

$$E\left(\sum_{j=1}^{n} |X_j|\right)^r < \sum_{j=1}^{n} E|X_j|^r \quad \text{for } 0 < r < 1.$$

Proof. Noting that for $r \geqslant 1$, $|X_j| \Big/ \left(\sum\limits_{k=1}^{n} |X_k|^r\right)^{1/r} \leqslant 1$ for each $j = 1, \cdots, n$, we have

$$\sum_{j=1}^{n} |X_j| \Big/ \left(\sum_{k=1}^{n} |X_k|^r\right)^{1/r} \geqslant \sum_{j=1}^{n} |X_j|^r \Big/ \left(\sum_{k=1}^{n} |X_k|^r\right) = 1,$$

which implies the first inequality. The second inequality can be seen from the fact that the above inequality has a inverse direction if $0 < r < 1$.

9.3 The Case $1 \leqslant r \leqslant 2$

9.3.a (von Bahr-Esseen). Let $1 \leqslant r \leqslant 2$ and $D_r \equiv (13.52/(2.6r)^r)\Gamma(r) \sin(r\pi/2) < 1$. If $X_1, \cdots, X_n$ are independent r.v.'s with $EX_j = 0, j = 1, \cdots, n$, then

$$E|S_n|^r \leqslant (1 - D_r)^{-1} \sum_{j=1}^{n} E|X_j|^r.$$

Proof. By simple calculations it follows that for any real number a,

$$|1 - \mathrm{e}^{\mathrm{i}a} + \mathrm{i}a| \leqslant 1.3|a|, \qquad |1 - \mathrm{e}^{\mathrm{i}a} + \mathrm{i}a| \leqslant 0.5a^2.$$

Multiplying the $(2-r)$th power of the first inequality by the $(r-1)$th power of the second one, we have $|1 - \mathrm{e}^{\mathrm{i}a} + \mathrm{i}a| \leqslant (3.38/(2.6)^r)|a|^r$. Let X be a r.v. with d.f. $F(x)$, c.f. $f(t)$, $EX = 0$ and $\beta_r \equiv E|X|^r < \infty$. Then

$$\begin{aligned} |1 - f(t)| &= \left| \int (1 - \mathrm{e}^{\mathrm{i}tx} + \mathrm{i}tx)\mathrm{d}F(x) \right| \\ &\leqslant (3.38/(2.6)^r)\beta_r|t|^r, \qquad -\infty < t < \infty,\ 1 \leqslant r \leqslant 2, \end{aligned}$$

which implies that

$$\begin{aligned} J &\equiv \int \frac{|1 - f(t)|^2}{|t|^{r+1}} \mathrm{d}t \\ &\leqslant 2\left(\frac{3.38}{(2.6)^r}\beta_r\right)^2 \int_0^b \frac{t^{2r}}{t^{r+1}} \mathrm{d}t + 2\int_b^\infty \frac{4}{t^{r+1}} \mathrm{d}t \\ &= \frac{2}{r}\left\{ \left(\frac{3.38}{(2.6)^r}\beta_r\right)^2 b^r + \frac{4}{b^r} \right\}. \end{aligned}$$

Choosing b so that this expression is a minimum, we obtain

$$J \leqslant (27.04/(2.6r)^r)\beta_r. \tag{43}$$

Then, by (36) we conclude that

$$E|X|^r \leqslant (2(1 - D_r))^{-1} E|X - X'|^r, \tag{44}$$

where X' denotes an independent copy of X, i.e. X and X' are i.i.d.

Consider the random vector $(X_1, \cdots, X_n)$, and let $(X_1', \cdots, X_n')$ be its independent copy. Put $S_n' = \sum_{j=1}^{n} X_j'$. By (43), 9.1.b and 8.7.a, we obtain

$$\begin{aligned} E|S_n|^r &\leqslant (2(1-D_r))^{-1} E|S_n - S_n'|^r \\ &\leqslant (2(1-D_r))^{-1} \sum_{j=1}^n E|X_j - X_j'|^r \\ &\leqslant (1-D_r)^{-1} \sum_{j=1}^n E|X_j|^r. \end{aligned}$$

9.3.b (Chatterji). Let $1 \leqslant r \leqslant 2$. If $E(X_{j+1}|S_j) = 0$ a.s. for $j = 1, \cdots, n-1$ (in particular, if $X_1, \cdots, X_n$ is a martingale difference sequence), $E|X_j|^r < \infty$, $j = 1, \cdots, n$, then

$$E|S_n|^r \leqslant 2^{2-r} \sum_{j=1}^n E|X_j|^r.$$

Proof. The cases $r = 1, 2$ are trivial. Consider the case where $1 < r < 2$. Noting that $\alpha \equiv \sup_x \{|1+x|^r - 1 - rx\}/|x|^r$ is finite, we have the elementary inequality

$$|a+b|^r \leqslant |a|^r + r|a|^{r-1} \cdot \operatorname{sgn}(a) b + \alpha |b|^r$$

for real numbers a and b. Note also that $\alpha \geqslant 1$. Integrating the inequality with $a = S_{n-1}$ and $b = X_n$, we obtain

$$E|S_{n-1} + X_n|^r \leqslant E|S_{n-1}|^r + \alpha E|X_n|^r.$$

The required inequality follows by induction.

9.4 The Case $r \geqslant 2$

Let $r \geqslant 2$ and $X_1, \cdots, X_n$ be a martingale difference sequence. Then

$$E|S_n|^r \leqslant C_r n^{r/2-1} \sum_{j=1}^n E|X_j|^r, \tag{45}$$

where $C_r = (8(r-1)\max(1, 2^{r-3}))^r$. If $X_1, \cdots, X_n$ are independent r.v.'s with zero means, then C_r can be replaced by $C_r' = \frac{1}{2} r(r-1) \max(1, 2^{r-3})(1 + 2r^{-1} D_{2m}^{(r-2)/2m})$, where $D_{2m} = \sum_{j=1}^m j^{2m-1}/(j-1)!$, and the integer m satisfies $2m \leqslant r < 2m+2$.

Proof. Put $\gamma_{rn} = E|X_n|^r$, $\beta_{rn} = \frac{1}{n}\sum_{j=1}^{n}\gamma_{rj}$. The inequality clearly holds when $r = 2$ or $\beta_{rn} = \infty$. Suppose $r > 2$ and $\beta_{rn} < \infty$.

Consider the martingale difference case first. Write

$$r_0 = \sup\{\tilde{r} \in [2, r]; (45) \text{ is true for } \tilde{r}\}.$$

We first claim that (45) is true for r_0. In fact, suppose that $r_m \uparrow r_0$ such that (45) holds for each r_m. Then, since $|S_n|^{r_m}$ is bounded by the integrable function $1 + |S_n|^r$ for all m, we have

$$E|S_n|^{r_0} = \lim_{m\to\infty} E|S_n|^{r_m} \leqslant \lim_{m\to\infty} C_{r_m} n^{r_m/2}\beta_{r_m n} = C_{r_0} n^{r_0/2}\beta_{r_0 n}.$$

Next, we claim that $r_0 = r$. If not, i.e., $r_0 < r$. Then, we can choose $r_2 \in (r_0, r)$ and define $r_1 = r_0 + 2(1 - r_0/r_2)$. Then, we have $r_0 < r_1 < r_2 < r$. We may assume that r_2 is close to r_0 so that $r_2 \leqslant 2r_1$ and

$$\gamma_{r_2 n}^{1/r_2} \leqslant 2\gamma_{r_1 n}^{1/r_1}.$$

Now, we begin to derive a contradiction that (45) holds for r_1. By Taylor's expansion,

$$\begin{aligned} |S_n|^{r_1} = &|S_{n-1}|^{r_1} + r_1 \mathrm{sgn}(S_{n-1})|S_{n-1}|^{r_1-1}X_n \\ &+ \frac{1}{2} r_1(r_1 - 1)|S_{n-1} + \theta X_n|^{r_1-2}X_n^2, \end{aligned}$$

where $0 < \theta < 1$. Noting that

$$|S_{n-1} + \theta X_n|^{r_1-2} \leqslant \max(1, 2^{r_1-3})(|S_{n-1}|^{r_1-2} + |X_n|^{r_1-2}),$$

we obtain

$$\Delta_n \equiv E(|S_n|^{r_1} - |S_{n-1}|^{r_1}) \leqslant \frac{1}{2} r_1\delta_{r_1}(E|S_{n-1}|^{r_1-2}X_n^2) + \gamma_{r_1 n}), \quad (46)$$

where $\delta_{r_1} = (r_1 - 1)\max(1, 2^{r_1-3})$. By the Hölder inequality, we have

$$E(|S_{n-1}|^{r_1-2}X_n^2) \leqslant (E|S_{n-1}|^{r_0})^{(r_2-2)/r_2}(E|X_n|^{r_2})^{2/r_2}. \quad (47)$$

By the Lyapounov inequality 8.3.c,

$$\begin{aligned} \beta_{r_0 n}^{(r_2-2)/r_2} &= \left(\frac{1}{n}\sum_{j=1}^{n}\gamma_{r_0 n}\right)^{(r_2-2)/r_2} \leqslant \left(\frac{1}{n}\sum_{j=1}^{n}\gamma_{r_0 n}^{r_1/r_0}\right)^{r_0(r_2-2)/(r_1 r_2)} \\ &\leqslant \left(\frac{1}{n}\sum_{j=1}^{n}\gamma_{r_1 n}\right)^{r_0(r_2-2)/(r_1 r_2)} = \beta_{r_1 n}^{(r_1-2)/r_1}. \end{aligned}$$

By (46) and the hypothesis that (45) holds for r_0, it follows that

$$\begin{aligned}E(|S_{n-1}|^{r_1-2}X_n^2) &\leqslant (C_{r_0}\beta_{r_0,n-1}(n-1)^{r_0/2})^{(r_2-2)/r_2}\gamma_{r_2n}^{2/r_2}\\ &\leqslant C_{r_0}^{(r_2-2)/r_2}\beta_{r_1,n-1}^{(r_1-2)/r_1}(n-1)^{(r_1-2)/2}\gamma_{r_2n}^{2/r_2}\\ &\leqslant C_{r_0}^{(r_2-2)/r_2}\beta_{r_1,n-1}^{(r_1-2)/r_1}(n-1)^{(r_1-2)/2}4\gamma_{r_1n}^{2/r_1}.\end{aligned}$$

Since $C_r = (8\delta_r)^r$, C_r is increasing in r and $C_r > 1$, hence

$$C_{r_0}^{(r_2-2)/r_2} \leqslant C_{r_1}^{(r_2-2)/r_2} = C_{r_1}C_{r_1}^{-2/r_2} \leqslant C_{r_1}C_{r_1}^{-1/r_1} = C_{r_1}(8\delta_{r_1})^{-1}.$$

Therefore we obtain

$$E(|S_{n-1}|^{r_1-2}X_n^2) \leqslant C_{r_1}(2\delta_{r_1})^{-1}(n-1)^{(r_1-2)/2}\beta_{r_1,n-1}^{(r_1-2)/r_1}\gamma_{r_1n}^{2/r_1}.$$

Thus

$$\Delta_n \leqslant \frac{1}{2}r_1\delta_{r_1}\{C_{r_1}(2\delta_{r_1})^{-1}(n-1)^{(r_1-2)/2}\beta_{r_1,n-1}^{(r_1-2)/r_1}\gamma_{r_1n}^{2/r_1} + \gamma_{r_1n}\}.$$

We need the following fact: let $y_1, \cdots, y_n$ be non-negative numbers. Put $z_n = (y_1 + \cdots + y_n)/n$, then, for all $x \geqslant 1$,

$$\sum_{j=2}^{n}(j-1)^{x-1}z_{j-1}^{(x-1)/x}y_j^{1/x} \leqslant n^x x^{-1} z_n, \tag{48}$$

see Dharmadhikari et al, (1968). Using this fact we obtain

$$\begin{aligned}E|S_n|^{r_1} &= \sum_{j=1}^{n}\Delta_j\\ &\leqslant \frac{1}{2}r_1\delta_{r_1}\{C_{r_1}(2\delta_{r_1})^{-1}\cdot 2r_1^{-1}n^{r_1/2}\beta_{r_1n} + n\beta_{r_1n}\}\\ &= \frac{1}{2}r_1\delta_{r_1}\{(r_1\delta_{r_1})^{-1}C_{r_1}n^{r_1/2} + n\}\beta_{r_1n}.\end{aligned}$$

Noting that $(r_1\delta_{r_1})^{-1}C_{r_1} > 1$ and $n^{r_1/2} \geqslant n$, we see that the second term in the brackets on the RHS of the above inequality is smaller than the first. Therefore

$$E|S_n|^{r_1} \leqslant \frac{1}{2}r_1\delta_{r_1}2(r_1\delta_{r_1})^{-1}C_{r_1}n^{r_1/2}\beta_{r_1n} = C_{r_1}n^{r_1/2}\beta_{r_1n}.$$

Now the proof is complete for the martingale difference case.

Next suppose that $X_1, \cdots, X_n$ are independent. We also need only consider the case $r > 2$. Let m be the integer such that $r - 2 < 2m \leqslant r$.

For $1 \leqslant p \leqslant 2m$, let A_p denote the set of all p-tuples $k = (k_1, \cdots, k_p)$ such that the k_i's are positive integers satisfying $k_1 + \cdots + k_p = 2m$. Let

$$T(i_1, \cdots, i_p) = \sum (2m)!/(k_1! \cdots k_p!) E(X_{i_1}^{k_1} \cdots X_{i_p}^{k_p}),$$

where the summation is over $k \in A_p$. Then

$$ES_n^{2m} = \sum_{p=1}^{2m} \sum{}^{*} T(i_1, \cdots, i_p),$$

where $\sum^*$ denotes summation over the region $1 \leqslant i_1 < \cdots < i_p \leqslant n$. If $p > m$ and $k \in A_p$, then $\min(k_1, \cdots, k_p) = 1$. Thus $p > m \Rightarrow T(i_1, \cdots, i_p) = 0$. Moreover by Hölder's inequality

$$|E(X_{i_1}^{k_1} \cdots X_{i_p}^{k_p})| \leqslant \gamma_{2m,i_1}^{k_1/2m} \cdots \gamma_{2m,i_p}^{k_p/2m}.$$

Therefore

$$\begin{aligned}
ES_n^{2m} &\leqslant \sum_{p=1}^{m} \sum{}^{*} (\gamma_{2m,i_1}^{1/2m} + \cdots + \gamma_{2m,i_p}^{1/2m})^{2m} \\
&\leqslant \sum_{p=1}^{m} p^{2m-1} \sum{}^{*} (\gamma_{2m,i_1} + \cdots + \gamma_{2m,i_p}) \\
&= \sum_{p=1}^{m} p^{2m-1} \binom{n-1}{p-1} \sum_{j=1}^{n} \gamma_{2m,j} \\
&\leqslant \sum_{p=1}^{m} p^{2m-1} \left(\frac{n^{p-1}}{(p-1)!} \right) n\beta_{2m,n} \\
&\leqslant D_{2m} n^m \beta_{2m,n},
\end{aligned}$$

which implies

$$E|S_{n-1}|^{r-2} \leqslant (ES_{n-1}^{2m})^{(r-2)/2m} \leqslant D_{2m}^{(r-2)/2m} (n-1)^{(r-2)/2} \beta_{2m,n-1}^{(r-2)/2m}.$$

Noting that $\beta_{2m,n-1} \leqslant \beta_{r,n-1}^{2m/r}$ and $\gamma_{2n} \leqslant \gamma_{rn}^{2/r}$, from (46) we obtain

$$\begin{aligned}
\Delta_n &= \frac{1}{2} r\delta_r (\gamma_{2n} E|S_{n-1}|^{r-2} + \gamma_{rn}) \\
&\leqslant \frac{1}{2} r\delta_r \{ D_{2m}^{(r-2)/2m} (n-1)^{(r-2)/2} \beta_{r,n-1}^{(r-2)/r} \gamma_{rn}^{2/r} + \gamma_{rn} \}.
\end{aligned}$$

Hence, by (48),

$$E|S_n|^r = \sum_{j=1}^{n} \Delta_j \leqslant \frac{1}{2} r\delta_r (D_{2m}^{(r-2)/2m} 2r^{-1} n^{r/2} \beta_{rn} + n\beta_{rn})$$
$$\leqslant C_r' n^{r/2} \beta_{rn},$$

as desired.

9.5 Jack-knifed Variance

Let $X_1, \cdots, X_n$ be independent r.v.'s, $S = S(X_1, \cdots, X_n)$ be a statistic having finite second moment. Put $S_{(i)} = S(X_1, \cdots, X_{i-1}, X_{i+1}, \cdots, X_n)$ and $S_{(\cdot)} = \sum\limits_{i=1}^{n} S_{(i)}/n$. Then

$$E \sum_{i=1}^{n} (S_{(i)} - S_{(\cdot)})^2 \geqslant \frac{1}{n} \sum_{i=1}^{n} \text{Var} S_{(i)} \geqslant \frac{n}{n-1} \text{Var} S_{(\cdot)}. \tag{49}$$

Proof. Put $\mu = ES$, $A_i = E(S|X_i) - \mu$ and

$$A_{ij} = E(S|X_i, X_j) - A_i - A_j + \mu$$

etc. Then we have the following ANOVA decomposition:

$$S(X_1, \cdots, X_n) = \mu + \sum_{i} A_i + \sum_{i<j} A_{ij}$$
$$+ \sum_{i<j<k} A_{ijk} + \cdots + H(X_1, \cdots, X_n), \tag{50}$$

where all $2^n - 1$ r.v.'s on the right side have mean zero and are mutually uncorrelated. In fact, the coefficient of μ on the right side is

$$1 - \binom{n}{1} + \binom{n}{2} - \cdots = (1-1)^n = 0.$$

Likewise the coefficient of A_i is $(1-1)^{n-1} = 0$, the coefficient of A_{ij} is $(1-1)^{n-2} = 0$, etc. The last term $H(X_1, \cdots, X_n)$, itself has first term $S(X_1, \cdots, X_n)$, which is the only term not canceling out. This verifies (50).

First assume $\mu_i \equiv ES_{(i)} = 0$, $i = 1, 2, \cdots, n$. Put $D = (n - 1) \sum\limits_{i=1}^{n} \text{Var} S_{(i)} - n^2 \text{Var} S_{(\cdot)}$ and let I, II, III be the three terms, from left to right, in (49). We have

$$\mathrm{I} - \mathrm{II} = D/n \quad \text{and} \quad \mathrm{II} - \mathrm{III} = D/(n(n-1)).$$

We now show that $D \geqslant 0$. From (50) for $S_{(i)}$ with $\mu_i = 0$ we can write

$$S_{(i)} = \sum_J S_{iJ},$$

where J indexes the $2^n - 2$ nonempty proper subsets of $\{1, 2, \cdots, n\}$. For example, with $i = 1$ and $J = \{2, 3\}$, $S_{1J} = A_{23}$ in (50) for $S_{(1)}$. The r.v.'s S_{iJ} satisfy

(i) $ES_{iJ} = 0$;

(ii) $S_{iJ} = 0$ if $i \in J$;

(iii) $ES_{iJ}S_{i'J'} = 0$ if $J \neq J'$. Define $S_{+J} = \sum_i S_{iJ}$ and notice that $ES_{+J}S_{+J'} = 0$ for $J \neq J'$. Therefore $En^2 S_{(\cdot)}^2 = E\sum_J S_{+J}^2$, and likewise $E(n-1)\sum_i S_{(i)}^2 = E\sum_J \{(n-1)\sum_i S_{iJ}^2\}$, so

$$D = E\sum_J \left\{(n-1)\sum_i S_{iJ}^2 - S_{+J}^2\right\}.$$

Letting n_J to be the number of elements in J and $\bar{S_J} = S_{+J}/(n - n_J)$, we obtain

$$D = E\sum_J \left\{(n_J - 1)\sum_i S_{iJ}^2 + (n - n_J)\sum_{i \notin J}(S_{iJ} - \bar{S_J})^2\right\} \geqslant 0.$$

Now if we drop the assumption that $\mu_i = 0$, II and III are unchanged, while I is increased by the amount $\sum(\mu_i - \mu_.)^2$, $\mu_. = \sum \mu_i/n$. This completes the proof of the inequality.

See Efron and Stein (1981).

9.6 Khintchine Inequality

Let $X_1, \cdots, X_n$ be i.i.d. r.v.'s with $P(X_1 = 1) = P(X_1 = -1) = \frac{1}{2}$ and $b_1, \cdots, b_n$ be any real numbers. Then for any $r > 0$, there exist constants $0 < C_r \leqslant C_r' < \infty$ such that

$$C_r\left(\sum_{j=1}^n b_j^2\right)^{r/2} \leqslant E\left|\sum_{j=1}^n b_j X_j\right|^r \leqslant C_r'\left(\sum_{j=1}^n b_j^2\right)^{r/2}.$$

Proof. At first, suppose that $r = 2k$, where k is an integer. Then, putting $T_n = \sum_{j=1}^{n} b_j X_j$, we have

$$ET_n^{2k} = \sum A_{l_1,\cdots,l_j} b_{i_1}^{l_1} \cdots b_{i_j}^{l_j} EX_{i_1}^{l_1} \cdots X_{i_j}^{l_j},$$

where $l_1, \cdots, l_j$ are positive integers with $\sum_{u=1}^{j} l_u = 2k$, $A_{l_1,\cdots,l_j} = (l_1 + \cdots + l_j)!/l_1! \cdots l_j!$, and $1 \leqslant i_1 < \cdots < i_j \leqslant n$. Since $EX_{i_1}^{l_1} \cdots X_{i_j}^{l_j} = 1$ when $l_1, \cdots, l_j$ are all even and 0 otherwise,

$$ET_n^{2k} = \sum A_{2p_1,\cdots,2p_j} b_{i_1}^{2p_1} \cdots b_{i_j}^{2p_j},$$

where $p_1, \cdots, p_j$ are positive integers with $\sum_{u=1}^{j} p_u = k$. Hence

$$ET_n^{2k} = \sum \frac{A_{2p_1,\cdots,2p_j}}{A_{p_1,\cdots,p_j}} A_{p_1,\cdots,p_j} b_{i_1}^{2p_1} \cdots b_{i_j}^{2p_j} \leqslant c_{2k} s_n^{2k},$$

where $s_n^2 = \sum_{j=1}^{n} b_j^2$ and

$$\begin{aligned} c_{2k} &= \sup \frac{A_{2p_1,\cdots,2p_j}}{A_{p_1,\cdots,p_j}} = \sup \frac{(2k)!}{(2p_1)! \cdots (2p_j)!} \frac{p_1! \cdots p_j!}{k!} \\ &\leqslant \sup \frac{2k(2k-1)\cdots(k+1)}{\prod_{u=1}^{j} 2p_u(2p_u - 1)\cdots(p_u+1)} \leqslant \frac{2k(2k-1)\cdots(k+1)}{2^{p_1+\cdots+p_j}} \\ &= \frac{2k(2k-1)\cdots(k+1)}{2^k} \leqslant k^k. \end{aligned}$$

Thus, when $r = 2k$ the upper bound is established with $C_{2k}' \leqslant k^k$. For $r \leqslant 2k$, we have

$$(E|T_n|^r)^{1/r} \leqslant (ET_n^{2k})^{1/2k} \leqslant c_{2k}^{1/2k} s_n,$$

hence the upper bound is obtained with $C_r' \leqslant k^{r/2}$, where k is the smallest integer $\geqslant r/2$.

To complete the proof of the Khintchine inequality, it suffices to establish the lower bound for $0 < r < 2$ since otherwise the Khintchine inequality follows from the fact that $(E|T_n|^r)^{1/r} \geqslant (ET_n^2)^{1/2}$. Take $r_1, r_2 > 0$ such that $r_1 + r_2 = 1$, $rr_1 + 4r_2 = 2$. By the Hölder inequality

$$s_n^2 = ET_n^2 \leqslant (E|T_n|^r)^{r_1} (ET_n^4)^{r_2} \leqslant (E|T_n|^r)^{r_1} (2^{1/2} s_n)^{4r_2}.$$

Here we have used the property that $f(r) = \log E|X|^r$ is a convex function on $[0, \infty)$, that can be proved by the Hölder inequality. The above inequality implies that

$$(E|T_n|^r)^{r_1} \geqslant 4^{-r_2} s_n^{2-4r_2} = 4^{-r_2} s_n^{rr_1},$$

$$E|T_n|^r \geqslant 4^{-r_2/r_1} s_n^r,$$

i.e., the lower bound is obtained for $0 < r < 2$ with $C_r \geqslant 4^{-r_2/r_1} = 2^{-(2-r)}$ and for $r \geqslant 2$ with $C_r \geqslant 1$.

9.7 Marcinkiewicz-Zygmund-Burkholder Type Inequalities

9.7.a (Marcinkiewicz-Zygmund-Burkholder inequality). Let $r \geqslant 1$, $X_1, X_2, \cdots$ be independent r.v.'s with $EX_n = 0$, $n = 1, 2, \cdots$, then there are positive constants $a_r \leqslant b_r$ such that

$$a_r E\left(\sum_{j=1}^n X_j^2\right)^{r/2} \leqslant E|S_n|^r \leqslant b_r E\left(\sum_{j=1}^n X_j^2\right)^{r/2},$$

$$a_r E\left(\sum_{j=1}^\infty X_j^2\right)^{r/2} \leqslant \sup_{n\geqslant 1} E|S_n|^r \leqslant b_r E\left(\sum_{j=1}^\infty X_j^2\right)^{r/2}.$$

Proof. By 9.1.a and 8.6, $E|S_n|^r < \infty \Longleftrightarrow E|X_j|^r < \infty$, $j = 1, \cdots, n \Longleftrightarrow E\left(\sum_{j=1}^n X_j^2\right)^{1/2} < \infty$. Hence the latter may be assumed to be true. Let $\{X_n', n \geqslant 1\}$ be i.i.d. with. $\{X_n, n \geqslant 1\}$ and $X_n^* = X_n - X_n'$. Moreover, let $\{V_n, n \geqslant 1\}$ be a sequence of i.i.d.r.v.'s with $P(V_1 = 1) = P(V_1 = -1) = \frac{1}{2}$, which is independent of $\{X_n, X_n', n \geqslant 1\}$. Since

$$E\left\{\sum_{j=1}^n V_j X_j^* | V_1, \cdots, V_n, X_1, \cdots, X_n\right\} = \sum_{j=1}^n V_j X_j,$$

it follows that for any integer $n \geqslant 1$, $\left\{\sum_{j=1}^n V_j X_j, \sum_{j=1}^n V_j X_j^*\right\}$ is a two-term martingale leading to the first inequality of

$$\begin{aligned} E\left|\sum_{j=1}^{n} V_j X_j\right|^r &\leqslant E\left|\sum_{j=1}^{n} V_j X_j^*\right|^r \\ &\leqslant 2^{r-1} E\left\{\left|\sum_{j=1}^{n} V_j X_j\right|^r + \left|\sum_{j=1}^{n} V_j X_j'\right|^r\right\} \\ &= 2^r E\left|\sum_{j=1}^{n} V_j X_j\right|^r . \end{aligned} \tag{51}$$

Applying Khintchine's inequality 9.6 to $E\left\{\left|\sum\limits_{j=1}^{n} V_j X_j\right|^r \middle| X_1, X_2, \cdots\right\}$, we obtain

$$C_r E\left(\sum_{j=1}^{n} X_j^2\right)^{r/2} \leqslant E\left|\sum_{j=1}^{n} V_j X_j\right|^r \leqslant C_r' E\left(\sum_{j=1}^{n} X_j^2\right)^{r/2},$$

which, in conjunction with (51), yields

$$C_r E\left(\sum_{j=1}^{n} X_j^2\right)^{r/2} \leqslant E\left|\sum_{j=1}^{n} V_j X_j^*\right|^r \leqslant 2^r C_r' E\left(\sum_{j=1}^{n} X_j^2\right)^{r/2} . \tag{52}$$

By the symmetry of X_j^*,

$$E\left|\sum_{j=1}^{n} V_j X_j^*\right|^r = E\left|\sum_{j=1}^{n} X_j^*\right|^r \leqslant 2^r E\left|\sum_{j=1}^{n} X_j\right|^r .$$

On the other hand, by Section 8.6,

$$E\left|\sum_{j=1}^{n} V_j X_j^*\right|^r = E\left|\sum_{j=1}^{n} X_j^*\right|^r \geqslant E\left|\sum_{j=1}^{n} X_j\right|^r .$$

Inserting these two inequalities into (52) yields the first desired inequality, which implies the second one immediately.

Burkholder (1973) extended the inequalities to the case where $(X_1, X_2, \cdots)$ is a martingale difference sequence and $r > 1$. The proof is omitted.

9.7.b (Rosenthal). If $\{X_k\}$ are independent non-negative, then for $r \geqslant 1$,

$$E\left(\sum_{k=1}^{n} X_k\right)^r \leqslant K_r \left(\left(\sum_{k=1}^{n} EX_k\right)^r + \sum_{k=1}^{n} EX_k^r\right).$$

Proof. If $r = 1$, the equality holds for $K_r = 1/2$. For the general case, we need only consider the case where the number of r.v.'s are finite since the case of infinitely many r.v.'s can be obtained by the monotone convergence theorem and making the number tend to infinity. At first, if $1 < r \leqslant 2$, then by 9.7.a, we have

$$\begin{aligned}
E\left(\sum_{k=1}^{n} X_k\right)^r &\leqslant 2^{r-1}\left(\left(\sum_{k=1}^{n} EX_k\right)^r + E\left(\sum_{k=1}^{n}(X_k - EX_k)\right)^r\right) \\
&\leqslant 2^{r-1}\left(\left(\sum_{k=1}^{n} EX_k\right)^r + K_r E\left(\sum_{k=1}^{n}(X_k - EX_k)^2\right)^{r/2}\right) \\
&\leqslant 2^{r-1}\left(\left(\sum_{k=1}^{n} EX_k\right)^r + K_r \left(\sum_{k=1}^{n} E(X_k - EX_k)^r\right)\right) \\
&\leqslant K_r \left(\left(\sum_{k=1}^{n} EX_k\right)^r + \sum_{k=1}^{n} EX_k^r\right).
\end{aligned}$$

Now, let us assume that the inequality holds for all $1 < r \leqslant 2^p$ and some integer p. Consider the case $2^p < r \leqslant 2^{p+1}$. Then, we have

$$\begin{aligned}
&E\left(\sum_{k=1}^{n} X_k\right)^r \leqslant 2^{r-1}\left(\left(\sum_{k=1}^{n} EX_k\right)^r + E\left(\sum_{k=1}^{n}(X_k - EX_k)\right)^r\right) \\
&\leqslant 2^{r-1}\left(\left(\sum_{k=1}^{n} EX_k\right)^r + K_r E\left(\sum_{k=1}^{n}(X_k - EX_k)^2\right)^{r/2}\right) \\
&\leqslant 2^{r-1}\Bigg(\left(\sum_{k=1}^{n} EX_k\right)^r + K_r\Bigg(\left(\sum_{k=1}^{n} E(X_k - EX_k)^2\right)^{r/2} \\
&\quad + \sum_{k=1}^{n} E(X_k - EX_k)^r\Bigg)\Bigg) \\
&\leqslant K_r \left(\left(\sum_{k=1}^{n} EX_k\right)^r + \left(\sum_{k=1}^{n} EX_k^2\right)^{r/2} + \sum_{k=1}^{n} EX_k^r\right).
\end{aligned}$$

Here, the last step follows by using the Hölder inequality and

$$\sum_{k=1}^{n} EX_k^2 \leqslant \left(\sum_{k=1}^{n} EX_k\right)^{(r-2)/(r-1)} \left(\sum_{k=1}^{n} EX_k^r\right)^{1/(r-1)}$$
$$\leqslant \max\left\{\left(\sum_{k=1}^{n} EX_k\right)^2, \left(\sum_{k=1}^{n} EX_k^r\right)^{2/r}\right\}.$$

9.7.c (Rosenthal). If $\{X_k\}$ is a sequence of independent r.v.'s whose expectations are 0, then for any $r \geqslant 2$,

$$E\left|\sum X_k\right|^r \leqslant C_r\left(\left(\sum EX_k^2\right)^{r/2} + \sum E|X_k|^r\right).$$

Proof. From 9.7.a and 9.7.b, we have

$$E\left|\sum X_k\right|^r \leqslant b_r E\left(\sum X_k^2\right)^{r/2} \leqslant b_r K_r\left(\left(\sum EX_k^2\right)^{r/2} + \sum E|X_k|^r\right).$$

9.8 Skorokhod Inequalities

Let $X_1, \cdots, X_n$ be independent r.v.'s with $|X_j| \leqslant c$ a.s., $j = 1, \cdots, n$, for some $c > 0$.

9.8.a. If the constants α and x are such that $4e^{2\alpha(x+c)}P(|S_n| \geqslant x) < 1$, then

$$E\exp(\alpha|S_n|) \leqslant e^{3\alpha x}/\{1 - 4e^{2\alpha(x+c)}P(|S_n| \geqslant x))(1 - P(|S_n| \geqslant x))\}. \tag{53}$$

Proof. Let $(X_1', \cdots, X_n')$ be a random vector which is i.i.d. with $(X_1, \cdots, X_n)$, and $X_j^* = X_j - X_j'$, $j = 1, \cdots, n$. We have $|X_j^*| \leqslant 2c$ a.s., $j = 1, \cdots, n$. Put $S_n' = \sum_{j=1}^{n} X_j'$ and $S_n^* = \sum_{j=1}^{n} X_j^*$. Clearly

$$P(|S_n^*| \geqslant 2x) \leqslant P(|S_n| \geqslant x) + P(|S_n'| \geqslant x) = 2P(|S_n| \geqslant x).$$

Hence we have $2e^{2\alpha(x+c)}P(|S_n^*| \geqslant 2x) < 1$. Then

$$E\exp(\alpha|S_n^*|) \leqslant e^{2\alpha x}/(1 - 2e^{2\alpha(x+c)}P(|S_n^*| \geqslant 2x))$$
$$\leqslant e^{2\alpha x}/(1 - 4e^{2\alpha(x+c)}P(|S_n| \geqslant x)).$$

Noting that $|S_n| - |S_n'| \leqslant |S_n^*|$, we have

$$Ee^{\alpha|S_n|}e^{-\alpha|S_n'|} \leqslant e^{2\alpha x}/(1 - 4e^{2\alpha(x+c)}P(|S_n| \geqslant x));$$

and in addition

$$Ee^{-\alpha|S'_n|} \geqslant e^{-\alpha x}(1 - P(|S_n| \geqslant x)).$$

The two last inequalities imply the desired conclusion.

9.8.b. If $P(|S_n| \geqslant x) \leqslant 1/(8e)$, then there exists a constant $C > 0$ such that

$$E|S_n|^m \leqslant Cm!(2x + 2C)^m.$$

Proof. Putting $\alpha = \frac{1}{2(x+c)}$ and making use of 9.8.a, we obtain

$$\begin{aligned} E\exp(\alpha|S_n|) &\leqslant e^{\frac{3x}{2(x+c)}}/((1 - 1/2)(1 - 1/(8e)) \\ &\leqslant 2e^{3/2}/(1 - 1/(8e)) \equiv C. \end{aligned}$$

Therefore, $E(\alpha^m|S_n|^m/m!) \leqslant C$, as desired.

9.9 Moments of Weighted Sums

Let $X_1, \cdots, X_n$ be i.i.d. r.v.'s with $EX_1 = 0$, $E|X_1|^{2p+\gamma} < \infty$ for some integer $p > 0$ and $0 \leqslant \gamma < 2, a_1, \cdots, a_n$ be numbers satisfying $\sum\limits_{j=1}^{n} a_j^2 = 1$. Then, when $\gamma = 0$,

$$E\left(\sum_{j=1}^{n} a_j X_j\right)^2 = EX_1^2, \qquad p = 1,$$

$$E\left(\sum_{j=1}^{n} a_j X_j\right)^{2p} < \left(\frac{3}{2}\right)^{p-2} (2p-1)!!EX^{2p}, \qquad p \geqslant 2;$$

when $0 < \gamma < 2$,

$$\begin{aligned} E\left|\sum_{j=1}^{n} a_j X_j\right|^{2+\gamma} < &\left\{1+2\Gamma(3+\gamma)\frac{1}{\pi}\left(\frac{2^{1-\gamma}}{\Gamma(3+\gamma)}+\frac{2}{\gamma}+\frac{3}{16(2-\gamma)}\right)\sin\frac{\gamma}{2}\pi\right\} \\ &\cdot E|X_1|^{2+\gamma}, \quad p = 1, \end{aligned}$$

$$\begin{aligned} E\left|\sum_{j=1}^{n} a_j X_j\right|^{2p+\gamma} < &\left\{1+2\Gamma(2p+1+\gamma)\frac{1}{\pi}\left(\frac{2^{1-\gamma}}{\Gamma(2p+1+\gamma)}+\frac{2}{\gamma(2p)!}\right.\right. \\ &\left.\left.+\frac{1}{(2-\gamma)(2p+2)!!}\left(\frac{3}{2}\right)^p+\frac{2}{\gamma(2p)!!}\left(\frac{3}{2}\right)^{p-2}\right)\sin\frac{\gamma}{2}\pi\right\} \\ &\cdot E|X_1|^{2p+\gamma}, \quad p \geqslant 2. \end{aligned}$$

The proof can be found in Tao and Cheng (1981).

9.10 Doob Crossing Inequalities

Let $(Y_1, Y_2, \cdots, Y_n)$ be a sequence of real numbers, $S_i = \sum_{j=1}^{i} Y_j$ and $a < b$ be two real numbers. Define $\tau_1 = \inf\{j \leqslant n; S_j < a\}$. By induction, define $\zeta_k = \inf\{j \in (\tau_k, n]; S_j > b\}$ and $\tau_{k+1} = \inf\{j \in (\zeta_k, n]; S_j < a\}$, where we use the convention that $\inf\{j \in \varnothing\} = n + 1$. Note that both the τ's and ζ's are stopping times if the sequence (Y) is random. We call the largest k such that $\zeta_k \leqslant n$ the number of upcrossings over the interval $[a, b]$. Similarly, we define the number of downcrossing of the sequence (Y) over the interval $[a, b]$ by the number of upcrossings of the sequence $(-Y)$ over the interval $[-b, -a]$.

9.10.a (upcrossing inequality). Let $\{Y_j, \mathscr{A}_j,\ 1 \leqslant j \leqslant n\}$ be a submartingale, $a < b$ be real numbers. Put $\nu_{ab}^{(n)}$ as the number of upcrossings of $[a, b]$ by $(Y_1, \cdots, Y_n)$. Then

$$E\nu_{ab}^{(n)} \leqslant \frac{1}{b-a}(E(Y_n - a)^+ - E(Y_1 - a)^+) \leqslant \frac{1}{b-a}(EY_n^+ + |a|).$$

Proof. Consider first the case where $Y_j \geqslant 0$ for any j and $0 = a < b$. When $n = 2$, the inequality is confirmed by

$$\begin{aligned} E(b\nu_{ab}^{(2)}) + EY_1 &= \int_{\{Y_1=0, Y_2 \geqslant b\}} b\mathrm{d}P + \int_{\{Y_1>0\}} Y_1 \mathrm{d}P \\ &\leqslant \int_{\{Y_1=0, Y_2 \geqslant b\}} Y_2 \mathrm{d}P + \int_{\{Y_1>0\}} Y_2 \mathrm{d}P \leqslant EY_2. \end{aligned}$$

Suppose inductively that the inequality holds with n replaced by $n - 1$ for all submartingales, and put

$$Z_j = Y_j, \qquad 1 \leqslant j \leqslant n-2,$$

$$Z_{n-1} = \begin{cases} Y_n, & \text{if } 0 < Y_{n-1} < b, \\ Y_{n-1}, & \text{otherwise.} \end{cases}$$

For $A \in \mathscr{A}_{n-2}$,

$$\begin{aligned} \int_A Z_{n-2} \mathrm{d}P &= \int_A Y_{n-2} \mathrm{d}P \leqslant \int_A Y_{n-1} \mathrm{d}P \\ &\leqslant \int_{A\{Y_{n-1} \geqslant b\}} Y_{n-1} \mathrm{d}P + \int_{A\{0<Y_{n-1}<b\}} Y_n \mathrm{d}P = \int_A Z_{n-1} \mathrm{d}P. \end{aligned}$$

Hence $E(Z_{n-1}|\mathscr{A}_{n-2}) \geqslant Z_{n-2}$ a.s. Clearly, for $2 \leqslant j \leqslant n-1$, $E(Z_j|\mathscr{A}_{j-1}) \geqslant Z_{j-1}$ a.s. and so $\{Z_j, \mathscr{A}_j, 1 \leqslant j \leqslant n-1\}$ is a nonnegative submartingale.

Let $\bar{\nu}_{ab}^{(n)}$ be the number of upcrossings of $[0, b]$ by $(Z_1, \cdots, Z_{n-1})$. Then

$$\nu_{ab}^{(n)} = \bar{\nu}_{ab}^{(n)} + I(Y_{n-1} = 0, Y_n \geqslant b).$$

Hence by the inductive hypothesis

$$\begin{aligned} E(b\nu_{ab}^{(n)}) + EY_1 &= E(b\bar{\nu}_{ab}^{(n)}) + E(bI(Y_{n-1} = 0, Y_n \geqslant b)) + EY_1 \\ &\leqslant EZ_{n-1} + E(bI(Y_{n-1} = 0, Y_n \geqslant b)) \\ &\leqslant \int_{\{0<Y_{n-1}<b\}} Y_n \mathrm{d}P + \int_{\{Y_{n-1} \geqslant b\}} Y_{n-1} \mathrm{d}P \\ &\quad + \int_{\{Y_{n-1}=0, Y_n \geqslant b\}} Y_n \mathrm{d}P \\ &\leqslant \int_{\{Y_{n-1}>0\}} Y_n \mathrm{d}P + \int_{\{Y_{n-1}=0, Y_n \geqslant b\}} Y_n \mathrm{d}P \leqslant EY_n. \end{aligned}$$

In the general case we apply the case just proved to $\{(Y_j - a)^+, 1 \leqslant j \leqslant n\}$ which is a nonnegative submartingale and $\nu_{ab}^{(n)}$ is the number of upcrossings of $[0, b-a]$ by $((Y_1 - a)^+, \cdots, (Y_n - a)^+)$.

9.10.b (downcrossing inequality). Let $\{Y_j, \mathscr{A}_j, 1 \leqslant j \leqslant n\}$ be a supermartingale, $a < b$ be real numbers. Put $\mu_{ab}^{(n)}$ as the number of downcrossings of $[a, b]$ by $(Y_1, \cdots, Y_n)$. Then

$$E\mu_{ab}^{(n)} \leqslant \frac{1}{b-a}(E(Y_1 \wedge b) - E(Y_n - b)).$$

Proof. $\{-Y_j, 1 \leqslant j \leqslant n\}$ is a submartingale and $\mu_{ab}^{(n)}$ is $\nu_{-b,-a}^{(n)}$ for this submartingale. Hence the upcrossing inequality becomes

$$E\mu_{ab}^{(n)} \leqslant \frac{1}{-a-(-b)} E\{(-Y_n + b)^+ - (-Y_1 + b)^+\} = \frac{1}{b-a}\{(b-Y_n)^+ - (b-Y_1)^+\}.$$

Substituting $(b-x)^+ = b - (b \wedge x)$ into the above, this is the same as in the desired inequality.

9.11 Moments of Maximal Partial Sums

Let $X_1, \cdots, X_n$ be independent r.v.'s, $r > 0$. Put

$$x_0 = \inf\left\{x > 0 : P\{\max_{1 \leqslant j \leqslant n} |S_j| \geqslant x\} \leqslant (2 \cdot 4^r)^{-1}\right\}.$$

Then

$$E \max_{1\leqslant j\leqslant n} |S_j|^r \leqslant 2\cdot 4^r E \max_{1\leqslant j\leqslant n} |X_j|^r + 2(4x_0)^r.$$

If, moreover, the X_j's are symmetric, and $x_0 = \inf\{x > 0 : P(|S_n| \geqslant x) \leqslant (8\cdot 3^r)^{-1}\}$, then

$$E|S_n|^r \leqslant 2\cdot 3^r E \max_{1\leqslant j\leqslant n} |X_j|^r + 2(3x_0)^r.$$

Proof. We only show the latter using the second inequality in 5.10. The proof of the former is similar by using the first inequality in 5.10. By integration by parts and the second inequality in 5.10, for u satisfying $4\cdot 3^r P(|S_n| \geqslant u) \leqslant 1/2$, we have

$$\begin{aligned}
E|S_n|^r &= 3^r \left(\int_0^u + \int_u^\infty\right) P(|S_n| \geqslant 3x)\mathrm{d}x^r \\
&\leqslant (3u)^r + 4\cdot 3^r \int_u^\infty (P(|S_n|\geqslant x))^2 \mathrm{d}x^r + 3^r \int_u^\infty P(\max_{1\leqslant j\leqslant n} |X_j|\geqslant x)\mathrm{d}x^r \\
&\leqslant (3u)^r + 4\cdot 3^r P(|S_n|\geqslant u) \int_0^\infty P(|S_n| \geqslant x)\mathrm{d}x^r + 3^r E \max_{1\leqslant j\leqslant n} |X_j|^r \\
&\leqslant 2(3u)^r + 2\cdot 3^r E \max_{1\leqslant j\leqslant n} |X_j|^r.
\end{aligned}$$

Since this holds for arbitrary $u > x_0$, the second inequality is shown.

9.12 Doob Inequalities

9.12.a. Let $\{Y_n, \mathscr{A}_n, n \geqslant 1\}$ be a nonnegative submartingale. Then

$$E\left\{\max_{1\leqslant j\leqslant n} Y_j\right\} \leqslant \frac{\mathrm{e}}{\mathrm{e}-1}(1 + E(Y_n \log^+ Y_n)),$$

$$E\left\{\sup_{n\geqslant 1} Y_n\right\} \leqslant \frac{\mathrm{e}}{\mathrm{e}-1}\left(1 + \sup_{n\geqslant 1} E(Y_n \log^+ Y_n)\right),$$

and for $p > 1$,

$$E\left\{\max_{1\leqslant j\leqslant n} Y_j^p\right\} \leqslant \left(\frac{p}{p-1}\right)^p EY_n^p,$$

$$E\left\{\sup_{n\geqslant 1} Y_n^p\right\} \leqslant \left(\frac{p}{p-1}\right)^p \sup_{n\geqslant 1} EY_n^p.$$

Proof. We shall prove the first and third inequalities, the second and the fourth are consequences of the first and the third respectively. Put

$Y_n^* = \max_{1\leqslant j\leqslant n} Y_j$. For a r.v. $X \geqslant 0$ with $EX^p < \infty$ and d.f. $F(x)$, using integration by parts we have

$$EX^p = p\int_0^\infty t^{p-1}(1-F(t))\mathrm{d}t, \quad p>0. \tag{54}$$

Hence, and using the inequality 6.5.a, we obtain

$$\begin{aligned} EY_n^* - 1 \leqslant E(Y_n^*-1)^+ &= \int_0^\infty P(Y_n^*-1 \geqslant x)\mathrm{d}x \\ &\leqslant \int_0^\infty \frac{1}{x+1}\int_{\{Y_n^*\geqslant x+1\}} Y_n \mathrm{d}P\mathrm{d}x \\ &= EY_n \int_0^{Y_n^*-1} \frac{\mathrm{d}x}{x+1} = EY_n \log Y_n^*. \end{aligned}$$

We need the following elementary inequality. For constants $a \geqslant 0$ and $b > 0$, $a\log b \leqslant a\log^+ a + b\mathrm{e}^{-1}$. Set $g(b) = a\log^+ a + b\mathrm{e}^{-1} - a\log b$. Then $g''(b) = a/b^2 > 0$ and $g'(a\mathrm{e}) = \mathrm{e}^{-1} - a/(a\mathrm{e}) = 0$. Thus, $g(a\mathrm{e}) = a\log^+ a - a\log a \geqslant 0$ is the minimum of $g(b)$. The inequality is proved. Applying this inequality, we obtain

$$EY_n^* - 1 \leqslant EY_n \log^+ Y_n + \mathrm{e}^{-1}EY_n^*,$$

from which the first inequality is immediate.

If $p > 1$, using (54), the inequality 6.5.a and the Hölder inequality, we obtain

$$\begin{aligned} EY_n^{*p} &= p\int_0^\infty x^{p-1}P(Y_n^* \geqslant x)\mathrm{d}x \\ &\leqslant p\int_0^\infty x^{p-2}\int_{\{Y_n^*\geqslant x\}} Y_n\mathrm{d}P\mathrm{d}x \\ &= pEY_n\int_0^{Y_n^*} x^{p-2}\mathrm{d}x \\ &= \frac{p}{p-1}EY_n(Y_n^*)^{p-1} \\ &\leqslant \frac{p}{p-1}(EY_n^p)^{1/p}(EY_n^{*p})^{(p-1)/p}, \end{aligned}$$

which yields the third inequality.

9.12.b. Let $X_1, \cdots, X_n$ be i.i.d. r.v.'s with $E|X_j|^r < \infty$ for some $r, j = 1, \cdots, n$. Then

$$E\max_{1\leqslant j\leqslant n}|S_j|^r \leqslant 8\max_{1\leqslant j\leqslant n}E|S_j|^r, \quad r>0,$$

$$E \max_{1\leqslant j\leqslant n} |S_j|^r \leqslant 8E|S_n|^r, \quad r \geqslant 1.$$

Proof. Since $\{|S_j|^r, j = 1, \ldots, n\}$ is a non-negative submartingale when $r \geqslant 1$, $\max_{1\leqslant j\leqslant n} E|S_j|^r = E|S_n|^r$, and hence the two desired inequalities are the same if $r \geqslant 1$. By integrating the Lévy inequality 5.4.b, it follows that

$$E \max_{1\leqslant j\leqslant n} |S_j - m(S_j - S_n)|^r \leqslant 2E|S_n|^r.$$

By Markov's inequality 6.2.d, $P\{|S_j| \geqslant (2E|S_j|^r)^{1/r}\} \leqslant 1/2$, which implies $|m(S_j)|^r \leqslant 2E|S_j|^r$. Therefore, using the c_r-inequality 9.1.a, we obtain

$$\begin{aligned} E \max_{1\leqslant j\leqslant n} |S_j|^r \leqslant & c_r E \max_{1\leqslant j\leqslant n} |S_j - m(S_j - S_n)|^r + c_r \max_{1\leqslant j\leqslant n} |m(S_j - S_n)|^r \\ \leqslant & 2c_r E|S_n|^r + 2c_r \max_{1\leqslant j\leqslant n} E|S_j|^r \leqslant 4c_r \max_{1\leqslant j\leqslant n} E|S_j|^r. \end{aligned}$$

Since $4c_r \leqslant 8$ if $0 < r \leqslant 2$, the desired inequality is proved. If $r > 2$, noting that $\left(\frac{r}{r-1}\right)^r \leqslant 8$, the third inequality of 9.12.a implies the conclusion.

9.13 Equivalence Conditions for Moments

Let $\{X_n, n \geqslant 1\}$ be a sequence of i.i.d. r.v.'s.

9.13.a (Marcinkiewicz-Zygmund-Burkholder). The following statements are equivalent:

(1) $E(|X_1| \log^+ |X_1|) < \infty$ for $r = 1$ or $E|X_1|^r < \infty$ for $r > 1$;

(2) $E\left(\sup_{n\geqslant 1} |S_n/n|^r\right) < \infty$ for $r \geqslant 1$;

(3) $E\left(\sup_{n\geqslant 1} |X_n|/n|^r\right) < \infty$ for $r \geqslant 1$.

Proof. For any integer n, define $\mathscr{F}_{n+1-j} = \sigma\{S_j/j, X_{j+1}, X_{j+2}, \cdots\}$. Then, $\{\mathscr{F}_1, \cdots, \mathscr{F}_n\}$ is a sequence of increasing σ fields and with respect to it, $\{S_n/n, \cdots, S_2/2, S_1\}$ forms a martingale. Hence, $\{|S_n|/n, \cdots, |S_2|/2, |S_1|\}$ is a nonnegative submartingale.

Applying Doob's inequality 9.12, it follows that

$$E \max_{j\leqslant n} |S_j|/j \leqslant \frac{\mathrm{e}}{\mathrm{e}-1} E(|X_1| \log^+ |X_1|),$$

and for $r > 1$,

$$E \max_{j\leqslant n} |S_j/j|^r \leqslant \left(\frac{r}{r-1}\right)^r E(|X_1|^r).$$

Making $n \to \infty$ yields (1)⇒ (2) and consequently (2)⇒(3) by noting that $X_n = S_n - S_{n-1}$.

When $r > 1$, from $\sup_{n\geqslant 1} |X_n/n|^r \geqslant |X_1|^r$, we conclude (3)⇒(1). When $r = 1$, (3) implies that

$$\begin{aligned}\infty > & \int_0^\infty P\left(\sup_{n\geqslant 1}|X_n|/n \geqslant x\right)\mathrm{d}x \\ = & \int_0^\infty \left[1 - \prod_{n\geqslant 1}(1 - P(|X_n|/n \geqslant x))\right]\mathrm{d}x \\ \geqslant & \int_0^\infty \left[1 - \exp\left(-\sum_{n\geqslant 1} P(|X_1|/n \geqslant x)\right)\right]\mathrm{d}x.\end{aligned}$$

Choose an M such that for any $x \geqslant M$,

$$\sum_{n\geqslant 1} P(|X_n|/n \geqslant x) < 1.$$

Then, we have

$$\begin{aligned}\infty > & \int_M^\infty \sum_{n\geqslant 1} P(|X_1|/n \geqslant x)\mathrm{d}x \\ = & \sum_{n\geqslant 1} EI(|X_1| \geqslant Mn)\int_M^{|X_1|/n}\mathrm{d}x \\ = & \sum_{n\geqslant 1} EI(|X_1| \geqslant Mn)[|X_1|/n - M] \\ = & EI(|X_1| \geqslant M) \sum_{|X_1|/M\geqslant n\geqslant 1}[|X_1|/n - M] \\ \geqslant & \frac{1}{2}EI(|X_1| \geqslant M)[|X_1|\log(|X_1|/M) - |X_1|],\end{aligned}$$

which implies (1) for $r = 1$. The proof is now complete.

9.13.b (Siegmund-Teicher). Suppose $EX_1 = 0$. Then the following statements are equivalent:

(i) $E(X_1^r L(|X_1|)/L_2(|X_1|)) < \infty$ for $r = 2$ or $E|X_1|^r < \infty$ for $r > 2$;

(ii) $E\left(\sup_{n\geqslant 1} |S_n/\sqrt{nL_2(n)}|^r\right) < \infty$ for $r \geqslant 2$;

(iii) $E\left(\sup_{n\geqslant 1}|X_n/\sqrt{nL_2(n)}|^r\right)<\infty$ for $r\geqslant 2$,

where $L(x)=1\vee\log x, L_2(x)=L(L(x))$.

Proof. First consider the situation when $r=2$. We will prove that (i)⇒(ii)⇒(iii)⇒(i). Put

$$a_n=(nL_2(n))^{-1},\quad b_n=n^{1/2}(L_2(n))^{-1},$$

Define

$$X_n^{'}=X_nI(|X_n|\leqslant b_n),\quad X_n^{''}=X_n-X_n^{'};\quad S_n^{'}=\sum_{j=1}^n X_j^{'},\quad S_n^{''}=S_n-S_n^{'}.$$

To prove (i)⇒(ii), assume that the d.f. of X_1 is symmetric and $EX_1^2=1$. From (i) we have

$$\begin{aligned}\sum_{j=1}^\infty a_jEX_j^{''2}&=\sum_{j=1}^\infty a_j\sum_{k=j}^\infty\int_{\{b_k<|X_1|\leqslant b_{k+1}\}}X_1^2\mathrm{d}P\\&=\sum_{k=1}^\infty\sum_{j=1}^k a_j\int_{\{b_k<|X_1|\leqslant b_{k+1}\}}X_1^2\mathrm{d}P\\&\leqslant c_1\sum_{k=1}^\infty(L(k)/L_2(k))\int_{\{b_k<|X_1|\leqslant b_{k+1}\}}X_1^2\mathrm{d}P\\&\leqslant c_1E\{X_1^2L(|X_1|)/L_2(|X_1|)\}<\infty,\end{aligned}$$

here $c_1,c_2,\cdots$ stand for positive constants. Similarly

$$\sum_{j=1}^n a_j^{1/2}E|X_j^{''}|\leqslant c_2EX_1^2<\infty.$$

Therefore

$$\begin{aligned}E\left(\sup_{n\geqslant 1}a_nS_n^{''2}\right)&\leqslant E\left(\sum_{j=1}^\infty a_j^{1/2}|X_j^{''}|\right)^2\\&\leqslant\sum_{j=1}^\infty a_jEX_j^{''2}+2\left(\sum_{j=1}^\infty a_j^{1/2}E|X_j^{''}|\right)^2<\infty.\quad(55)\end{aligned}$$

Then consider S_n'. Put $n_k = 3^k$. From the Lévy inequality 5.4.b, for any $x > 0$,

$$P\left\{\sup_{n\geqslant 1} a_n^{1/2}|S_n'| > x\right\} \leqslant \sum_{j=0}^{\infty} P\left\{a_{n_j}^{1/2} \sup_{n_j\leqslant n<n_{j+1}} |S_n'| > x\right\}$$

$$\leqslant 4\sum_{j=0}^{\infty} P\{a_{n_j}^{1/2} S_{n_{j+1}}' > x\}.$$

Using inequalities 6.1.a and 8.2, for $0 < t \leqslant b_{n_{j+1}}^{-1}, j = 0, 1, \cdots$,

$$P\{S_{n_{j+1}}' > a_{n_j}^{-1/2}x\} \leqslant \exp\{-ta_{n_j}^{1/2}x + t^2 n_{j+1}\}.$$

Putting $t = b_{n_{j+1}}^{-1}$, we have

$$\log P\{S_{n_{j+1}}' > a_{n_j}^{-1/2}x\} \leqslant -c_2(x - c_3)L_2(n_{j+1}).$$

Choosing x_0 such that $c_2(x_0 - c_3) \geqslant 2$, we have

$$\int_{x_0}^{\infty} xP\left\{\sup_{n\geqslant 1} a_n^{1/2}|S_n'| > x\right\} \mathrm{d}x$$

$$\leqslant c_4 \sum_{j=1}^{\infty} \int_{x_0}^{\infty} x \exp\{-c_2(x - c_3)L_2(n_j)\} \,\mathrm{d}x$$

$$\leqslant c_4 \sum_{j=1}^{\infty} \int_{x_0}^{\infty} x \exp\{-c_2(x - c_3)\log j\} \exp\{-c_2(x - c_3)L_2(3)\} \,\mathrm{d}x$$

$$\leqslant c_4 \sum_{k=1}^{\infty} k^{-2} \int_{x_0}^{\infty} x \exp\{-c_2(x - c_3)L_2(3)\} \,\mathrm{d}x < \infty.$$

Thus we obtain

$$E\left\{\sup_{n\geqslant 1} a_n S_n'^2\right\} < \infty.$$

Combining this with (55), we find (ii) is true under the symmetric condition. Generally, let $X_1^*, X_2^*, \cdots$ be r.v.'s which are independent with $X_1, X_2, \cdots$ and having identical distribution with X_1. Put $S_n^* = \sum_{j=1}^{n} X_j^*$. Referring the result under the symmetric condition, we have

$$\begin{aligned}E\left\{\sup_{n\geqslant 1} a_n S_n^2\right\} &= E\left\{\sup_{n\geqslant 1} a_n|S_n - E(S_n^*|X_1, X_2, \cdots)|^2\right\}\\ &\leqslant E\left\{\sup_{n\geqslant 1} a_n E(|S_n - S_n^*|^2|X_1, X_2, \cdots)\right\}\\ &\leqslant E\left\{E(\sup_{n\geqslant 1} a_n|S_n - S_n^*|^2|X_1, X_2, \cdots)\right\}\\ &\leqslant E\left\{\sup_{n\geqslant 1} a_n|S_n - S_n^*|^2\right\} < \infty.\end{aligned}$$

To prove (ii)⇒(iii), we only need to focus on the following relation.

$$a_n X_n^2 = a_n(S_n - S_{n-1})^2 \leqslant 2(a_n S_n^2 + a_{n-1}S_{n-1}^2).$$

Now assume that (iii) is true, thus we have

$$\sum_{k=1}^{\infty} P\left\{\sup_{n\geqslant 1} a_n X_n^2 \geqslant k\right\} < \infty.$$

Let F be the d.f. of X_1^2. Without loss of generality, assume that $F(1) > 0$. Then

$$\begin{aligned}&\int_1^\infty \int_1^\infty (1 - F(xL_2(x)y))\,\mathrm{d}y\mathrm{d}x\\ &\leqslant \sum_{n=1}^{\infty}\sum_{j=1}^{\infty}(1 - F(a_n^{-1}j)) \leqslant -\sum_{j=1}^{\infty} \log \prod_{n=1}^{\infty} F(a_n^{-1}j)\\ &\leqslant c_5 \sum_{j=1}^{\infty}\left(1 - \prod_{n=1}^{\infty} F(a_n^{-1}j)\right) \leqslant c_6 \sum_{j=1}^{\infty} P\left\{\sup_{n\geqslant 1} a_n X_n^2 \geqslant j\right\} < \infty.\end{aligned}$$

Put $u = xL_2(x)y$. Then

$$\int_1^\infty \int_{xL_2(x)}^\infty (1 - F(u))\,\mathrm{d}u(xL_2(x))^{-1}\,\mathrm{d}x < \infty.$$

Denote the inverse of function $x \to xL_2(x)$ by φ. By the Fubini theorem, we have

$$\int_1^\infty \left\{\int_1^{\varphi(u)} (xL_2(x))^{-1}\,\mathrm{d}x\right\}(1 - F(u))\,\mathrm{d}u < \infty.$$

Since $\varphi(u) \sim (u/L_2(u))$ $(u \to \infty)$ and $\int_1^t (xL_2(x))^{-1}\,\mathrm{d}x \sim L(t)/L_2(t)$ $(t \to \infty)$, the expression above is equivalent to

$$\int_1^\infty (L(u)/L_2(u))(1 - F(u))\,\mathrm{d}u < \infty.$$

Furthermore, it is also equivalent to (i).

Now consider the condition when $r > 2$. Let $Y_1, Y_2, \cdots$ be a sequence of independent and non-negative r.v.'s. By the Lyapounov inequality 8.3.c and the Hölder inequality 8.3.a, for $0 < b < c < d$, we have

$$\sum_{j=1}^{\infty} a_j^c EY_j^c \leqslant \sum_{j=1}^{\infty} (a_j^b EY_j^b)^{(d-c)/(d-b)} (a_j^d EY_j^d)^{(c-b)/(d-b)}$$
$$\leqslant \left(\sum_{j=1}^{\infty} a_j^b EY_j^b\right)^{(d-c)/(d-b)} \left(\sum_{j=1}^{\infty} a_j^d EY_j^d\right)^{(c-b)/(d-b)}.$$

If the following condition is satisfied

$$\sum_{j=1}^{\infty} a_j^r EY_j^r < \infty \quad \text{and} \quad \sum_{j=1}^{\infty} a_j^\alpha EY_j^\alpha < \infty$$

(when r is an integer, $\alpha = 1$; otherwise $\alpha = r - [r]$), then for $\alpha \leqslant h \leqslant r$,

$$\sum_{j=1}^{\infty} a_j^h EY_j^h < \infty. \tag{56}$$

We will show that $E(\Sigma_{j=1}^{\infty} a_j EY_j)^r < \infty$ can be proved from the above formula. Just consider the condition when r is not an integer. Put $k = r - \alpha$, and using independence, we obtain

$$E\left(\sum_{j=1}^{\infty} a_j Y_j\right)^r = E\left(\sum_{j=1}^{\infty} a_j Y_j\right)^\alpha \left(\sum_{j=1}^{\infty} a_j Y_j\right)^k$$
$$\leqslant E\left(\sum_{j=1}^{\infty} a_j^\alpha Y_j^\alpha\right) \left\{\sum_{j=1}^{\infty} a_j^k Y_j^k + \cdots \right.$$
$$\left. + k! \sum_{1 \leqslant j_1 < \cdots < j_k} a_{j_1} Y_{j_1} \cdots a_{j_k} Y_{j_k}\right\}$$
$$= \sum_{j=1}^{\infty} a_j^r Y_j^r + \sum_{i \neq j} a_i^\alpha EY_i^\alpha a_j^k EY_j^k + \cdots$$
$$+ k! \sum_{1 \leqslant j_1 < \cdots < j_k, j \neq j_l, 1 \leqslant l \leqslant k} a_{j_1} EY_{j_1} \cdots a_{j_k} EY_{j_k} a_j^\alpha EY_j^\alpha$$
$$+ k! \sum_{1 \leqslant j_1 < \cdots < j_{k-1}, j \neq j_l, 1 \leqslant l \leqslant k} a_{j_1} EY_{j_1}$$
$$\cdots a_{j_{k-1}} EY_{j_{k-1}} a_j^{1+\alpha} EY_j^{1+\alpha}.$$

Each term on the RHS is limited by products like (56). For example, the last term is no more than

$$k!\left(\sum_{j=1}^{n}a_jEY_j\right)^{k-1}\left(\sum_{j=1}^{n}a_j^{\alpha+1}EY_j^{\alpha+1}\right).$$

To prove (i)⇒(ii). Introduce $X_n^{'}$, $X_n^{''}$, $S_n^{'}$ and $S_n^{''}$ as defined earlier, where $b_n = n^{1/r}$. For $h = \alpha$ or r,

$$\begin{aligned}\sum_{n=1}^{\infty}a_n^{h/2}E|X_n^{''}|^h &= \sum_{n=1}^{\infty}\sum_{j=n}^{\infty}\frac{1}{(nL_2(n))^{h/2}}\int_{\{b_j<|X_1|\leqslant b_{j+1}\}}|X_1|^h\,\mathrm{d}P\\ &\leqslant c_7\sum_{j=1}^{\infty}\frac{j^{1-h/2}}{(L_2(j))^{h/2}}\int_{\{b_j<|X_1|\leqslant b_{j+1}\}}|X_1|^h\,\mathrm{d}P\\ &= c_7\sum_{j=1}^{\infty}\frac{j^{h(1/r-1/2)}}{(L_2(j))^{h/2}}\int_{\{b_j<|X_1|\leqslant b_{j+1}\}}|X_1|^h\,\mathrm{d}P\\ &\leqslant c_8E|X_1|^r<\infty.\end{aligned}$$

Noting the result about $\{Y_n\}$ which has been proved, we obtain

$$E\sup_{n\geqslant 1}\frac{|S_n^{''}|^r}{(nL_2(n))^{r/2}}\leqslant E\left(\sup_{n\geqslant 1}a_n\sum_{j=1}^{n}|X_j^{''}|\right)^r\leqslant E\left(\sum_{n=1}^{\infty}a_n|X_n^{''}|\right)^r<\infty.$$

Then we will prove $E\sup_{n\geqslant 1}(|S_n^{'}|^r/(nL_2(n))^{r/2}) < \infty$, or equivalently prove: for sufficiently big x_0,

$$\int_{x_0}^{\infty}x^{r-1}P\left\{\sup_{n\geqslant 1}a_n^{1/2}|S_n^{'}|>x\right\}\mathrm{d}x<\infty.$$

The proof is similar to that under the condition when $r = 2$, except choosing $t = (L_2(n_{j+1})/n_{j+1})^{1/2}$ instead of $b_{n_{j+1}}^{-1}$. Without the symmetric assumption, the proof is also similar. Thus it is clear that (i)⇒(ii), (ii)⇒(i) and (ii)⇒(iii)⇒(i) under the present condition.

9.14 Serfling Inequalities

Let $S_{a,n} = X_{a+1} + \cdots + X_{a+n}$, $M_{a,n} = \max_{1\leqslant j\leqslant n}|S_{a,j}|$. The integer $a_0 > 0$ is arbitrary but fixed.

9.14.a. Let $r \geqslant 2$. Suppose that there exists a functional $g(a,n)$ of the joint d.f. of $X_{a+1}, \cdots, X_{n+n}$, satisfying

$$E|S_{a,n}|^r \leqslant g^{r/2}(a,n) \quad \text{for } a \geqslant a_0, n \geqslant 1,$$

$$g(a,k) + g(a+k,l) \leqslant g(a,k+l) \quad \text{for } a \geqslant a_0, 1 \leqslant k < k+l.$$

Then

$$EM_{a,n}^r \leqslant (\log_2 2n)^r g^{r/2}(a,n) \quad \text{for } a \geqslant a_0, n \geqslant 1.$$

Remark. As a consequence, if $X_{a+1}, \cdots, X_{a+n}$ are mutually orthogonal, then

$$EM_{a,n}^r \leqslant (\log_2 2n)^2 (\sigma_{a+1}^2 + \cdots + \sigma_{a+n}^2)^{r/2},$$

where $\sigma_j^2 = EX_j^2$.

Proof. Let $N > 1$ be a given integer and put $m = [(N+1)/2]$. For $m < n \leqslant N$ we have

$$\begin{aligned} S_{a,n}^2 &= S_{a,m}^2 + S_{a+m,n-m}^2 + 2S_{a,m}S_{a+m,n-m} \\ &\leqslant M_{a,m}^2 + M_{a+m,N-m}^2 + 2|S_{a,m}|M_{a+m,N-m}. \end{aligned}$$

For $1 \leqslant n \leqslant m$ we have $S_{a,n}^2 \leqslant M_{a,m}^2$. Therefore

$$M_{a,N}^2 \leqslant M_{a,m}^2 + M_{a+m,N-m}^2 + 2|S_{a,m}|M_{a+m,N-m},$$

and, by Minkowski's inequality 9.2.a,

$$\begin{aligned} (EM_{a,N}^r)^{2/r} \leqslant& (EM_{a,m}^r)^{2/r} + (EM_{a+m,N-m}^r)^{2/r} \\ &+2(E|S_{a,m}M_{a+m,N-m}|^{r/2})^{2/r}. \end{aligned}$$

Clearly, the desired conclusion is true for $N = 1$. Suppose inductively that it is true for $n < N$. Then, defining

$$f(k) = (\log_2 2k)^2, \qquad k \geqslant 1,$$

we have, using Cauchy-Schwarz's inequality 8.3.b,

$$\begin{aligned} (EM_{a,N}^r)^{2/r} \leqslant& f(m)g(a,m) + f(N-m)g(a+m,N-m) \\ &+2\{(E|S_{a,m}|^r)^{1/2}(E(M_{a+m,N-m}^r))^{1/2}\}^{2/r} \\ \leqslant& f(m)g(a,m) + f(N-m)g(a+m,N-m) \\ &+2(E|S_{a,m}|^r)^{1/r} f^{1/2}(N-m)g^{1/2}(a+m,N-m). \end{aligned}$$

Then, by the conditions on g, the inequality $2AB \leqslant A^2 + B^2$ and the fact that $f(N-m) \leqslant f(m)$, it follows that

$$(EM_{a,N}^r)^{2/r} \leqslant (f(m) + f^{1/2}(m))g(a,N). \tag{57}$$

Now note that

$$f(2k) = (\log_2 2k + 1)^2 \geqslant f(k) + 2f^{1/2}(k), \quad k \geqslant 1,$$

and since $2^{1/2}(2k-1) \geqslant 2k$ if $k \geqslant 2$,

$$f(2k-1) = (\log_2(2^{1/2}(2k-1)) + 1/2)^2 \geqslant f(k) + f^{1/2}(k), \quad k \geqslant 2.$$

Hence, whether $N = 2m$ or $N = 2m-1$,

$$f(N) \geqslant f(m) + f^{1/2}(m), \quad N > 1,$$

so that (57) yields

$$EM_{a,N}^r \leqslant (\log_2 2N)^r g^{r/2}(a,N).$$

This proves the conclusion by induction.

9.14.b. Let $r > 2$. Suppose that

$$E|S_{a,n}|^r \leqslant g^{r/2}(n) \quad \text{for } a \geqslant a_0, n \geqslant 1,$$

where $g(n)$ is a nondecreasing function satisfying $2g(n) \leqslant g(2n)$ and $g(n)/g(n+1) \to 1$ as $n \to \infty$. Then there exists a constant K (which may depend on r, g and the joint d.f. of X_j's) such that

$$EM_{a,n}^r \leqslant Kg^{r/2}(n) \quad \text{for } a \geqslant a_0, n \geqslant 1.$$

Proof. Let $k = r-1$ if r is an integer and $k = [r]$ otherwise. Put $\varepsilon = r - k$. It follows that the function

$$w(x) \equiv \sum_{j=1}^{k-1} \binom{k}{j} x^{-(j+\varepsilon)/r} + \sum_{j=1}^{k} \binom{k}{j} x^{-j/r} \downarrow 0 \quad \text{as} \quad x \to \infty.$$

Hence there exists x_0 such that

$$x \geqslant x_0 \Rightarrow w(x) \leqslant 2^{r\delta/2} - 2,$$

where $2/r < \delta < 1$. Also, since $g(n) \sim g(n+1)$, there exists n_0 such that

$$n \geqslant n_0 \Rightarrow g(n) \leqslant 2^{1-\delta} g(n-1).$$

By the hypothesis the quantity $q_n \equiv \sup_{a \geqslant a_0} EM_{a,n}^r / g^{r/2}(n)$ is finite. Define

$$K = \max\{q_1, q_2, \cdots, q_{n_0}, x_0\}.$$

Thus, for this K, the conclusion holds for all $n \leqslant n_0$. We shall show that it holds for all $N > n_0$ if it is assumed true for all $n < N$.

Let $N > n_0$ be given and put $m = [(N+1)/2]$. For $m < n \leqslant N$, we have

$$\begin{aligned}|S_{a,n}|^r &\leqslant (|S_{a,m}| + M_{a+m,N-m})^r \\ &\leqslant |S_{a,m}|^r + M_{a+m,N-m}^r + \sum_{j=0}^{k-1} \binom{k}{j} |S_{a,m}|^{j+\varepsilon} M_{a+m,N-m}^{k-j} \\ &\quad + \sum_{j=1}^{k} \binom{k}{j} |S_{a,m}|^j M_{a+m,N-m}^{k-j+\varepsilon}.\end{aligned}$$

For $1 \leqslant n \leqslant m$, we have $|S_{a,n}|^r \leqslant M_{a,m}^r$. It follows that

$$\begin{aligned}M_{a,N}^r \leqslant M_{a,m}^r + M_{a+m,N-m}^r &+ \sum_{j=0}^{k-1} \binom{k}{j} |S_{a,m}|^{j+\varepsilon} M_{a+m,N-m}^{k-j} \\ &+ \sum_{j=1}^{k} \binom{k}{j} |S_{a,m}|^j M_{a+m,N-m}^{k-j+\varepsilon}.\end{aligned} \tag{58}$$

Using Hölder's inequality 8.3.a, for $u \geqslant 0, v \geqslant 0$ with $u + v = r$,

$$\begin{aligned}E(|S_{a,m}|^u M_{a+m,N-m}^v) &\leqslant (E|S_{a,m}|^r)^{u/r} (EM_{a+m,N-m}^r)^{v/r} \\ &\leqslant K^{v/r} g^{u/2}(m) g^{v/2}(N-m) \leqslant K^{v/r} g^{r/2}(m)\end{aligned}$$

since $N - m \leqslant m$ and g is nondecreasing. Applying this result in each term on the RHS of (58) yields

$$EM_{a,N}^r \leqslant K g^{r/2}(m)(2 + w(K)).$$

Since $K \geqslant x_0, 2m \geqslant n_0$, the definitions of x_0 and n_0 and the assumptions on $g(\cdot)$ imply that

$$\begin{aligned}EM_{a,N}^r &\leqslant K 2^{r\delta/2} g^{r/2}(m) = K 2^{r(\delta-1)/2} (2g(m))^{r/2} \\ &\leqslant K 2^{r(\delta-1)/2} g^{r/2}(2m) \leqslant K g^{r/2}(2m-1) \\ &\leqslant K g^{r/2}(N),\end{aligned}$$

i.e., the conclusion holds for $n = N$. This completes the proof.

9.15 Average Fill Rate

Let $y_1, \cdots, y_n$ be i.i.d. positive r.v.'s and define $x_i = y_i \wedge t$ where t is a positive r.v. The expected fill rate over period n is defined by $\rho_n = E\frac{x_1+\cdots+x_n}{y_1+\cdots+y_n}$.

9.15.a. For any fixed n, $\rho_1 \geqslant \rho_n \geqslant \liminf_{m\to\infty}\rho_m$.

This result was proved by Chen et al (2003) who also conjectured that ρ_n is decreasing in n.

9.15.b. The sequence of fill rates ρ_n is decreasing in n.

This conjecture was proved by Banerjeea et al (2005). We shall further generalize it as the following result and provide a simple proof.

9.15.c. Let $y_1, \cdots, y_n$ be a set of positive, exchangeable r.v.'s and define $x_i = g(y_i, t)$, where $g(y,t)/y$ is non-increasing in y for each given t. Then $\rho_n = E\frac{x_1+\cdots+x_n}{y_1+\cdots+y_n}$ is deceasing in n.

Proof. Denote $s_x = x_1 + \cdots + x_n$, $s_y = y_1 + \cdots + y_n$, $n \geqslant 2$. Then, using

$$\frac{1}{s_y - y_i} = \frac{1}{s_y} + \frac{y_i}{s_y(s_y - y_i)},$$

we obtain

$$\begin{aligned}
\rho_{n-1} &= \frac{1}{n}\sum_{i=1}^{n} E\frac{s_x - x_i}{s_y - y_i} \quad (\text{ by exchangeability }) \\
&= \frac{1}{n}\sum_{i=1}^{n} E(s_x - x_i)\left[\frac{1}{s_y} + \frac{y_i}{s_y(s_y - y_i)}\right] \\
&= \rho_n - \frac{\rho_n}{n} + \frac{1}{n}\sum_{i=1}^{n} E(s_x - x_i)\frac{y_i}{s_y(s_y - y_i)} \\
&= \rho_n + \frac{1}{n}\sum_{i=1}^{n} E\left[\frac{s_x y_i^2}{s_y^2(s_y - y_i)} - \frac{x_i y_i}{s_y(s_y - y_i)}\right] \\
&= \rho_n + \frac{1}{n}\sum_{i=1}^{n}\sum_{j=1}^{n} E\frac{x_j y_i}{s_y^2}\left(\frac{y_i}{s_y - y_i} - \frac{y_j}{s_y - y_j}\right) \\
&= \rho_n + \frac{1}{n}\sum_{y_i > y_j} E\frac{x_j y_i - x_i y_j}{s_y^2}\left(\frac{y_i}{s_y - y_i} - \frac{y_j}{s_y - y_j}\right) \\
&> \rho_n,
\end{aligned}$$

where the last step follows from the fact that

$$\frac{y_i}{s_y - y_i} - \frac{y_j}{s_y - y_j} > 0$$

and $x_j y_i - x_i y_j = y_i y_j (x_j/y_j - x_i/y_i) > 0$ since $g(y,t)/y$ is non-increasing.

Refer to Chen, Lin and Thomas (2003).

References

Banerjeea A, Paul A. 2005. Average fill rate and horizon length. Operations Research Letters, 33: 525-530.

Burkholder D L. 1973. The 1971 Wald memorial lectures: distribution function inequalities for martingales. Ann. Probab., 1: 19-42.

Chatterji S D. 1969. An L^p-convergence theorem. Ann. Math. Statist., 40: 1068-1070.

Chen J, Lin D K, Thomas D J. 2003. On the single item fill rate for a finite horizon. Oper. Res. Lett., 31: 119-123.

Chow Y S, Teicher H. 1997. Probability Theory: Independence, Interchangeability, Martingales. 2nd ed., New York: Springer-Verlag.

Dharmadhikari S W, Fabian V, Jogdeo K. 1968. Bounds on the moments of martingales. Ann. Math. Statist., 39: 1719-1723.

Doob J L. 1953. Stochastic Processes. New York: Wiley.

Efron B, Stein C. 1981. The jackknife estimate of variance. Ann. Statist., 3: 586-596.

Loéve M. 1977. Probability Theory. 4th ed.. New York: Springer-Verlag.

McCabe B J, Shepp L A. 1970. On the supremum of S_n/n. Ann. Math. Statist., 41: 2166-2168.

Petrov V V. 1995. Limit Theorems of Probability Theory: Sequences of Independent Random Variables. Oxford & New York: Clarendon Press & Oxford University Press.

Rosenthal H P. 1970. On the subspaces of L^p $(p > 2)$ spanned by sequences of independent random variables. Israel J. Math., 8: 273-303.

Serfling R J. 1970a. Moment inequalities for the maximum cumulative sum. Ann. Math. Statist., 41: 1227-1234.

Serfling R J. 1970b. Convergence properties of S_n under moment restrictions. Ann. Math. Statist., 41: 1235-1248.

Shorack G R, Wellner J A. 1986. Empirical processes with applications to statistics. New York: Wiley.

Siegmund D. 1969. On moments of the maximum of normed partial sums. Ann. Math. Statist., 40: 527-531.

Teicher H. 1971. Completion of a dominated ergodic theorem. Ann. Math. Statist., 42: 2156-2158.

Tao B, Cheng P. 1981. An inequality of moments. Chin. Ann. Math., 2: 451-461 (in Chinese)

von Bahr B, Esseen C G. 1965. Inequalities for the r-th absolute moment of a sum of random variables, $1 \leqslant r \leqslant 2$. Ann. Math. Statist., 36: 299-303.

Chapter 10

Inequalities Related to Mixing Sequences

Most theorems in classical probability theory are derived under the assumption of independence of random variables or events. However, in many practical cases, the random variables are dependent. Thus, investigation on dependent random variables has both theoretical and practical importance. In chapter 6, we have introduced the concept of martingales that is a big class of dependent random variables. There is another class of dependent random variables, that is, time-dependent observations or time series. It is imaginable that observations at nearer time instants have stronger dependency while the dependency becomes weaker when the time distance increases. To describe such sequences of random variables, we shall introduce the concept of mixing. There are at least six different definitions of mixing sequences. In this chapter, we only give three most commonly used definitions.

Let $\{X_n, n \geqslant 1\}$ be a sequence of r.v.'s. Denote the σ-algebra $\mathscr{F}_a^b = \sigma(X_n, a \leqslant n \leqslant b), \mathbb{N} = \{1, 2, \cdots\}$. $L_p(\mathscr{F}_a^b)$ is a set of all $\mathscr{F}_a^b$-measurable r.v.'s with p-th moments.

A sequence $\{X_n, n \geqslant 1\}$ is said to be α-mixing (or strong mixing) if

$$\alpha(n) \equiv \sup_{k\in\mathbb{N}} \sup_{A\in\mathscr{F}_1^k, B\in\mathscr{F}_{k+n}^\infty} |P(AB) - P(A)P(B)| \to 0 \quad \text{as} \quad n \to \infty;$$

it is said to be ρ-mixing if

$$\rho(n) \equiv \sup_{k\in\mathbb{N}} \sup_{X\in L_2(\mathscr{F}_1^k), Y\in L_2(\mathscr{F}_{k+n}^\infty)} \frac{|EXY - EXEY|}{\sqrt{\mathrm{Var}X\mathrm{Var}Y}} \to 0 \quad \text{as} \quad n \to \infty;$$

it is said to be φ-mixing (or uniformly strong mixing) if

$$\varphi(n) \equiv \sup_{k\in\mathbb{N}} \sup_{A\in\mathscr{F}_1^k, B\in\mathscr{F}_{k+n}^\infty} |P(B|A) - P(B)| \to 0 \quad \text{as} \quad n \to \infty.$$

We have the relations φ-mixing $\Rightarrow$ ρ-mixing $\Rightarrow$ α-mixing.

10.1 Covariance Estimates for Mixing Sequences

10.1.a. Let $\{X_n, n \geqslant 1\}$ be an α-mixing sequence, $X \in \mathscr{F}_1^k, Y \in \mathscr{F}_{k+n}^\infty$ with $|X| \leqslant C_1, |Y| \leqslant C_2$ a.s. Then

$$|EXY - EXEY| \leqslant 4C_1C_2\alpha(n).$$

Proof. By the property of conditional expectation, we have

$$\begin{aligned} |EXY - EXEY| &= |E\{X(E(Y|\mathscr{F}_1^k) - EY)\}| \\ &\leqslant C_1E|E(Y|\mathscr{F}_1^k) - EY| \\ &= C_1|E\xi\{E(Y|\mathscr{F}_1^k) - EY\}|, \end{aligned}$$

where $\xi = \mathrm{sgn}(E(Y|\mathscr{F}_1^k) - EY) \in \mathscr{F}_1^k$, i.e.,

$$|EXY - EXEY| \leqslant C_1|E\xi Y - E\xi EY|.$$

With the same argument procedure it follows that

$$|E\xi Y - E\xi EY| \leqslant C_2|E\xi\eta - E\xi E\eta|,$$

where $\eta = \mathrm{sgn}(E(\xi|\mathscr{F}_{k+n}^\infty) - E\xi)$. Therefore

$$|EXY - EXEY| \leqslant C_1C_2|E\xi\eta - E\xi E\eta|. \tag{59}$$

Put $A = \{\xi = 1\}, B = \{\eta = 1\}$. It is clear that $A \in \mathscr{F}_1^k, B \in \mathscr{F}_{k+n}^\infty$. Using the definition of α-mixing, we obtain

$$\begin{aligned} |E\xi\eta - E\xi E\eta| &= |P(AB) + P(A^cB^c) - P(A^cB) - P(AB^c) \\ &\quad -(P(A) - P(A^c))(P(B) - P(B^c))| \\ &\leqslant 4\alpha(n). \end{aligned}$$

Inserting it into (59) yields the desired inequality.

10.1.b. Let $\{X_n, n \geqslant 1\}$ be an α-mixing sequence, $X \in L_p(\mathscr{F}_1^k), Y \in L_q(\mathscr{F}_{k+n}^\infty)$ with $p, q, r \geqslant 1$ and $\frac{1}{p} + \frac{1}{q} + \frac{1}{r} = 1$. Then

$$|EXY - EXEY| \leqslant 8\alpha(n)^{1/r}||X||_p||Y||_q,$$

where $||X||_p = (E|X|^p)^{1/p}$.

Proof. At first assume that $|Y| \leqslant C$ a.s. and $1 < p < \infty$. For some $a > 0$, define $X' = XI(|X| \leqslant a)$ and $X'' = X - X'$. Now, by 10.1.a,

$$\begin{aligned}|EXY - EXEY| &\leqslant |EX'Y - EX'EY| + |EX''Y - EX''EY| \\ &\leqslant 4Ca\alpha(n) + 2CE|X''|,\end{aligned}$$

where $E|X''| \leqslant a^{1-p}E|X|^p$. Putting $a = ||X||_p\alpha(n)^{-1/p}$ we obtain

$$|EXY - EXEY| \leqslant 6C||X||_p\alpha(n)^{1-1/p}.$$

If Y is not bounded a.s., put $Y' = YI(|Y| \leqslant b)$ and $Y'' = Y - Y'$ for some $b > 0$. Similarly

$$|EXY - EXEY| \leqslant 6b||X||_p\alpha(n)^{1-1/p} + 2||X||_p(E|Y''|^{\frac{qr}{q+r}})^{\frac{q+r}{qr}},$$

where

$$(E|Y''|^{\frac{qr}{q+r}})^{\frac{q+r}{qr}} \leqslant (b^{-q+\frac{qr}{q+r}}E|Y|^q)^{\frac{q+r}{qr}} = b^{-q/r}||Y||_q^{(q+r)/r}.$$

Putting $b = ||Y||_q\alpha(n)^{-1/q}$, we obtain the desired inequality.

10.1.c. Let $\{X_n, n \geqslant 1\}$ be a ρ-mixing sequence, $X \in L_p(\mathscr{F}_1^k), Y \in L_q(\mathscr{F}_{k+n}^\infty)$ with p, $q \geqslant 1$ and $\frac{1}{p} + \frac{1}{q} = 1$. Then

$$|EXY - EXEY| \leqslant 4\rho(n)^{\frac{2}{p}\wedge\frac{2}{q}}\|X\|_p\|Y\|_q.$$

Proof. Without loss of generality assume that $p \geqslant 2$, which implies that $q \leqslant 2$. Let $Y' = YI(|Y| \leqslant C)$ and $Y'' = Y - Y'$ for some $C > 0$. Write

$$|EXY - EXEY| \leqslant |EXY' - EXEY'| + |EXY'' - EXEY''|. \tag{60}$$

By the definition of ρ-mixing and Hölder's inequality,

$$|EXY' - EXEY'| \leqslant \rho(n)\|X\|_2\|Y\|_2 \leqslant \rho(n)C^{1-q/2}\|X\|_p\|Y\|_p^{q/2},$$

$$\begin{aligned}|EXY''| &\leqslant (E|Y''|^q)^{1-2/p}(E(|X|^{p/2}|Y''|^{q/2}))^{2/p} \\ &\leqslant (E|Y|^q)^{1-2/p}\Big(E(|X|^{p/2}E|Y''|^{q/2} \\ &\quad +\rho(n)(E|X|^p)^{1/2}(E|Y''|^q)^{1/2}\Big)^{2/p} \\ &\leqslant \|X\|_p\|Y\|_q^qC^{-q/p} + \rho(n)^{2/p}\|X\|_p\|Y\|_q\end{aligned}$$

and

$$|EXEY''| \leqslant \|X\|_p \|Y\|_q^q C^{-q/p}.$$

Inserting these estimates into (60) and taking $C = \|Y\|_q \rho(n)^{-2/q}$ yield the desired inequality.

10.1.d. Let $\{X_n, n \geqslant 1\}$ be a φ-mixing sequence, $X \in L_p(\mathscr{F}_1^k), Y \in L_q(\mathscr{F}_{k+n}^\infty)$ with $p, q \geqslant 1$ and $\frac{1}{p} + \frac{1}{q} = 1$. Then

$$|EXY - EXEY| \leqslant 2\varphi(n)^{1/p} \|X\|_p \|Y\|_q.$$

Proof. At first, we assume that X and Y are simple functions, i.e.,

$$X = \sum_i a_i I_{A_i}, \quad Y = \sum_j b_j I_{B_j},$$

where both $\sum_i$ and $\sum_j$ are finite sums and $A_i \bigcap A_r = \varnothing \ (i \neq r), B_j \bigcap B_l = \varnothing \ (j \neq l), A_i, A_r \in \mathscr{F}_1^k, B_j, B_l \in \mathscr{F}_{k+n}^\infty$. So

$$EXY - EXEY = \sum_{i,j} a_i b_j P(A_i B_j) - \sum_{i,j} a_i b_j P(A_i) P(B_j).$$

By Hölder's inequality we have

$$\begin{aligned}
&|EXY - EXEY| \\
&= \left| \sum_i a_i (P(A_i))^{1/p} \sum_j (P(B_j|A_i) - P(B_j)) b_j (P(A_i))^{1/q} \right| \\
&\leqslant \left(\sum_i |a_i|^p P(A_i) \right)^{1/p} \left(\sum_i P(A_i) \Big| \sum_j b_j (P(B_j|A_i) - P(B_j)) \Big|^q \right)^{1/q} \\
&\leqslant \|X\|_p \Bigg| \sum_i P(A_i) \Bigg(\sum_j |b_j|^q (P(B_j|A_i) \\
&\quad + P(B_j)) \Bigg) \Bigg(\sum_j |P(B_j|A_i) - P(B_j)| \Bigg)^{q/p} \Bigg|^{1/p} \\
&\leqslant 2^{1/q} \|X\|_p \|Y\|_q \max_i \left(\sum_j |P(B_j|A_i) - P(B_j)| \right)^{1/p}. \qquad (61)
\end{aligned}$$

Note that

$$\begin{aligned}\sum_j |P(B_j|A_i) - P(B_j)| &= \left(P\left(\overset{+}{\bigcup_j} B_j|A_i\right) - P\left(\overset{+}{\bigcup_j} B_j\right)\right) \\ &\quad - \left(P\left(\overset{-}{\bigcup_j} B_j|A_i\right) - P\left(\overset{-}{\bigcup_j} B_j\right)\right) \\ &\leqslant 2\varphi(n),\end{aligned}$$

where $\overset{+}{\bigcup_j}\left(\overset{-}{\bigcup_j}\right)$ is carried out over j such that $P(B_j|A_i) - P(B_j) > 0$ $(P(B_j|A_i) - P(B_j) < 0)$. Inserting it into (61) yields the desired estimate for the simple function case.

For the general case, let

$$X_N = \begin{cases} 0, & \text{if } |X| > N, \\ k/N, & \text{if } k/N < X \leqslant (k+1)/N, |X| \leqslant N; \end{cases}$$

$$Y_N = \begin{cases} 0, & \text{if } |Y| > N, \\ k/N, & \text{if } k/N < Y \leqslant (k+1)/N, |Y| \leqslant N. \end{cases}$$

Noting the result showed for X_N and Y_N and

$$E|X - X_N|^p \to 0, \quad E|Y - Y_N|^q \to 0 \quad \text{as} \quad N \to \infty,$$

we obtain the desired inequality.

10.2 Tail Probability on α-mixing Sequence

Let $\{X_n, n \geqslant 1\}$ be an α-mixing sequence. For any given integers p, q and k, let ξ_j be $\mathscr{F}_{(j-1)(p+q)+1}^{jp+(j-1)q}$ measurable, $j = 1, 2, \cdots, k$. Then for any $x > 0$,

$$P\left\{\max_{1\leqslant l\leqslant k} |\xi_1 + \cdots + \xi_l| > 2x\right\} \leqslant \frac{P\{|\xi_1 + \cdots + \xi_k| > x\} + k\alpha(q)}{\min\limits_{1\leqslant l\leqslant k-1} P\{|\xi_{l+1} + \cdots + \xi_k| \leqslant x\}}.$$

Proof. Let

$$\begin{aligned}&A = \{\max_{1\leqslant l\leqslant k} |\xi_1 + \cdots + \xi_l| > 2x\}, \quad B = \{|\xi_1 + \cdots + \xi_k| > x\}, \\ &A_1 = \{|\xi_1| > 2x\}, \\ &A_l = \{\max_{1\leqslant r\leqslant l-1} |\xi_1 + \cdots + \xi_r| \leqslant 2x\}, \quad |\xi_1 + \cdots + \xi_l| > 2x\}, \quad l = 2, \cdots, k, \\ &B_l = \{|\xi_{l+1} + \cdots + \xi_k| \leqslant x\}, \quad l = 1, \cdots, k-1, \quad B_k = \Omega.\end{aligned}$$

Then by α-mixing condition

$$\begin{aligned} P(A_l B_l) &\geqslant P(A_l)P(B_l) - \alpha(q) \\ &\geqslant P(A_l) \min_{1 \leqslant j \leqslant k-1} P(B_j) - \alpha(q), \quad l = 1, \cdots, k, \end{aligned}$$

and hence

$$\begin{aligned} P(B) &\geqslant \sum_{l=1}^{k} P(A_l B_l) \geqslant \sum_{l=1}^{k} P(A_l) \min_{1 \leqslant j \leqslant k-1} P(B_j) - k\alpha(q) \\ &= P(A) \min_{1 \leqslant l \leqslant k-1} P(B_l) - k\alpha(q), \end{aligned}$$

which implies the desired inequality.

See Lin (1982).

10.3 Estimates of 4-th Moment on ρ-mixing Sequence

Let $\{X_n, n \geqslant 1\}$ be a ρ-mixing stationary sequence with $EX_1 = 0$ and $EX_1^4 < \infty$. Then for any $\varepsilon > 0$ there exists a $C = C(\varepsilon, \rho(\cdot)) > 0$ such that for each $n \geqslant 1$,

$$ES_n^4 \leqslant C(n^{1+\varepsilon} EX_1^4 + \sigma_n^4),$$

where $\sigma_n^2 = ES_n^2$.

Proof. Denote $a_m = \|S_m\|_4$. It is clear that

$$a_{2m} \leqslant \|S_m + S_{k+m}(m)\|_4 + 2ka_1,$$

where $S_{k+m}(m) = \sum\limits_{j=k+m+1}^{k+2m} X_j$. Using the Cauchy-Schwarz inequality and by the definition of ρ-mixing, we have

$$\begin{aligned} &E|S_m + S_{k+m}(m)|^4 \\ &\leqslant 2a_m^4 + 6E|S_m S_{k+m}(m)|^2 + 8a_m^2 (E|S_m S_{k+m}(m)|^2)^{1/2} \\ &\leqslant 2a_m^4 + 6(\sigma_m^4 + \rho(k)a_m^4) + 8a_m^2(\sigma_m^4 + \rho(k)a_m^4)^{1/2} \\ &\leqslant 2(1 + 7\rho^{1/2}(k))a_m^4 + 8a_m^2\sigma_m^2 + 6\sigma_m^4 \\ &\leqslant (2^{1/4}(1 + 7\rho^{1/2}(k))^{1/4} a_m + 2\sigma_m)^4. \end{aligned}$$

It follows that

$$a_{2m} \leqslant 2^{1/4}(1 + 7\rho^{1/2}(k))^{1/4} a_m + 2\sigma_m + 2ka_1.$$

Let $0 < \varepsilon < 1/3$ and k be large enough such that $1 + 7\rho^{1/2}(k) \leqslant 2^{\varepsilon}$. By the recurrence method for each integer $r \geqslant 1$, we have

$$\begin{aligned} a(2^r) &\leqslant 2^{r(1+\varepsilon)/4} a_1 + 2\sum_{j=1}^{r} 2^{(j-1)(1+\varepsilon)/4}(\sigma(2^{r-j}) + ka_1) \\ &\leqslant c(2^{r(1+\varepsilon)/4} a_1 + \sigma(2^r)) \end{aligned}$$

for some $c > 0$, which implies the desired inequality.

Remark. Using the method in 10.4 and 10.5 below, we can extend the result to the non-stationary case.

See Peligrad (1987).

10.4 Estimates of Variances of Increments of ρ-mixing Sequence

Let $\{X_n, n \geqslant 1\}$ be a ρ-mixing sequence with $EX_n = 0$, $EX_n^2 < \infty$ for each $n \geqslant 1$. Then for any $\varepsilon > 0$, there exists a $C = C(\varepsilon) > 0$ such that

$$ES_k^2(n) \leqslant Cn \exp\left\{(1+\varepsilon)\sum_{j=0}^{[\log n]} \rho(2^j)\right\} \max_{k<j\leqslant k+n} EX_j^2 \tag{62}$$

for each $k \geqslant 1$ and $n \geqslant 1$, where $S_k(n) = \sum\limits_{j=k+1}^{k+n} X_j$.

Proof. Without loss of generality, assume $0 < \varepsilon < 1/4$. Let $m_j = [2^{j/(1+\varepsilon)}] + 1$. We shall prove that for some constant C_1 and any $n < 2^{N+1}$,

$$ES_k^2(n) \leqslant C_1 n \exp\left\{\sum_{j=1}^{N}(\rho(m_j) + 4m_j^{1/2}2^{-j/2})\right\} \max_{k<j\leqslant k+n} EX_j^2. \tag{63}$$

It is obvious that (63) holds for $n \leqslant 16$ by choosing $C_1 = 16$. Suppose that (63) holds for $n < 2^N$ with $C_1 = 16$, that is to say,

$$ES_k^2(n) \leqslant C_1 n \exp\left\{\sum_{j=1}^{N-1}(\rho(m_j) + 4m_j^{1/2}2^{-j/2})\right\} \max_{k<j\leqslant k+n} EX_j^2. \tag{64}$$

Now, we consider the case $2^N \leqslant n < 2^{N+1}$. Let $n_1 = [(n - m_N)/2]$ and $n_2 = n - m_N - n_1$. Then, both n_1 and n_2 are less than 2^N. By the induction assumption (64),

$$
\begin{aligned}
ES_k^2(n) = & ES_k^2(n_1) + ES_{k+n_1+m_N}^2(n_2) \\
& +2ES_k(n_1)S_{k+n_1+m_N}(n_2) + ES_{k+n_1}^2(m_N) \\
& +2ES_{k+n_1}(m_N)S_{k+n_1+m_N}(n_2) + 2ES_k(n_1)S_{k+n_1}(m_N) \\
\leqslant & (ES_k^2(n_1) + ES_{k+n_1+m_N}^2(n_2))(1+\rho(m_N)) + ES_{k+n_1}^2(m_N) \\
& +2(ES_{k+n_1}^2(m_N)ES_{k+n_1+m_N}^2(n_2))^{1/2} \\
& +2(ES_k^2(n_1)ES_{k+n_1}^2(m_N))^{1/2} \\
\leqslant & C_1 \exp\left\{\sum_{j=1}^{N-1}(\rho(m_j) + 4m_j^{1/2}2^{-j/2})\right\} \max_{k<j\leqslant k+n} EX_j^2 \\
& \cdot\left[(n_1+n_2)(1+\rho(m_N)) + m_N + 2\sqrt{n_1 m_N} + 2\sqrt{n_2 m_N}\right] \\
\leqslant & C_1 n \exp\left\{\sum_{j=1}^{N-1}(\rho(m_j) + 4m_j^{1/2}2^{-j/2})\right\} \max_{k<j\leqslant k+n} EX_j^2 \\
& \cdot\left[1+\rho(m_N) + 4(m_N)^{1/2}2^{-N/2}\right] \\
\leqslant & C_1 n \exp\left\{\sum_{j=1}^{N}(\rho(m_j) + 4m_j^{1/2}2^{-j/2})\right\} \max_{k<j\leqslant k+n} EX_j^2.
\end{aligned}
$$

This proves (63).

To finish the proof of (62), we first note that by the definition of m_j,

$$
\sum_{j=1}^{\infty} m_j^{1/2}2^{-j/2} < \infty.
$$

Finally, we need to estimate $\sum\limits_{j=1}^{N}\rho(m_j)$. We piecewise linearly extend $\rho(n)$ to all positive numbers, that is, $\rho(x) = (\rho_{i+1} - \rho_i)(x-i) + \rho_i$, if $i \leqslant x < i+1$, with $\rho_0 = 1$. Note that $\rho(x)$ is non-increasing. Therefore,

$$
\begin{aligned}
\sum_{j=1}^{N}\rho(m_j) \leqslant & \sum_{j=1}^{N}\rho(2^{j/(1+\varepsilon)}) \\
& \leqslant \int_0^N \rho(2^{x/(1+\varepsilon)})\mathrm{d}x = (1+\varepsilon)\int_0^{N/(1+\varepsilon)} \rho(2^x)\mathrm{d}x \\
& \leqslant (1+\varepsilon)\left(1 + \sum_{j=1}^{N}\rho(2^j)\right).
\end{aligned}
$$

The proof is complete.

Remark. The opposite inequality can also be shown. If, in addition,

$$ES_k^2(n)/\min_{k<j\leqslant k+n} EX_j^2 \to \infty \quad \text{as} \quad n\to\infty$$

uniformly in k and

$$\max_{k<j\leqslant k+n} EX_j^2 \leqslant a \min_{k<j\leqslant k+n} EX_j^2 \quad \text{for} \quad \text{some} \quad a\geqslant 1,$$

then, for any $\varepsilon>0$, there exist $C'=C'(\varepsilon,\rho(\cdot),a)>0$ and an integer N such that for each $k\geqslant 0$ and $n\geqslant N$,

$$ES_k^2(n)\geqslant C'n\exp\left\{-(1+\varepsilon)\sum_{j=0}^{[\log n]}\rho(2^j)\right\}\min_{k<j\leqslant k+n} EX_j^2. \tag{65}$$

See Shao (1995).

10.5 Bounds of $2+\delta$-th Moments of Increments of ρ-mixing Sequence

Let $\{X_n, n\geqslant 1\}$ be a ρ-mixing sequence with $EX_n=0$, $\sup_n E|X_n|^{2+\delta}<\infty$ for some $0<\delta<1$ and

$$\sum_{n=1}^{\infty}\rho(2^n)<\infty.$$

Then there exists a $C=C(\delta,\rho(\cdot))>0$ such that for each $n\geqslant 1$,

$$\begin{aligned}\sup_{k\geqslant 1} E|S_k(n)|^{2+\delta}\leqslant & C\{n^{1+\delta/2}(\sup_{k\geqslant 1} EX_k^2)^{1+\delta/2}\\ &+n\exp\{(C\log n)^{\delta/(2+\delta)}\}\sup_{k\geqslant 1}E|X_k|^{2+\delta}\}.\end{aligned} \tag{66}$$

Proof. Put $a_m=\sup_{k\geqslant 1}\|S_k(m)\|_{2+\delta}$, $\sigma_m=\sup_{k\geqslant 1}\|S_k(m)\|_2$ and $m_1=m+[m^{1/5}]$. Obviously

$$\|S_k(2m)\|_{2+\delta}\leqslant\|S_k(m)+S_{k+m_1}(m)\|_{2+\delta}+2m^{1/5}a_1.$$

It is not difficult to verify that for $x\geqslant 0$,

$$\begin{aligned}(1+x)^{2+\delta}&\leqslant 1+(2+\delta)^2(x+x^{1+\delta})+x^{2+\delta}\\ &\leqslant 1+9(x+x^{1+\delta})+x^{2+\delta}.\end{aligned}$$

Hence

$$E|S_k(m)+S_{k+m_1}(m)|^{2+\delta} \leqslant 2a_m^{2+\delta} + 9E|S_k(m)|^{1+\delta}|S_{k+m_1}(m)| \\ +9E|S_k(m)||S_{k+m_1}(m)|^{1+\delta}.$$

Moreover, by Hölder's inequality and 10.1.c we have

$$\begin{aligned} &E|S_k(m)|^{1+\delta}|S_{k+m_1}(m)| \\ &\leqslant \|S_k(m)\|_{2+\delta}^{\delta}\|S_k(m)S_{k+m_1}(m)\|_{(2+\delta)/2} \\ &\leqslant a_m^{\delta}\left\{\sigma_m^{2+\delta} + 4\rho([m^{1/5}])a_m^{2+\delta}\right\}^{2/(2+\delta)} \\ &\leqslant a_m^{\delta}\sigma_m^2 + 4\rho^{2/(2+\delta)}([m^{1/5}])a_m^{2+\delta}. \end{aligned}$$

Similarly

$$E|S_k(m)||S_{k+m_1}(m)|^{1+\delta} \leqslant a_m^{\delta}\sigma_m^2 + 4\rho^{2/(2+\delta)}([m^{1/5}])a_m^{2+\delta}.$$

Combining these inequalities yields

$$\begin{aligned} &E|S_k(m)+S_{k+m_1}(m)|^{2+\delta} \\ &\leqslant 2a_m^{2+\delta} + 18(a_m^{\delta}\sigma_m^2 + 4\rho^{2/(2+\delta)}([m^{1/5}])a_m^{2+\delta}) \\ &\leqslant \left\{\left[2(1+36\rho^{2/(2+\delta)}([m^{1/5}]))\right]^{1/(2+\delta)} a_m + 18\sigma_m\right\}^{2+\delta}, \end{aligned}$$

which implies that

$$a_{2m} \leqslant \left\{2(1+36\rho^{2/(2+\delta)}([m^{1/5}]))\right\}^{1/(2+\delta)} a_m + 18\sigma_m + 2m^{1/5}a_1.$$

Noting the condition on $\rho(\cdot)$, we have

$$\rho(n) \leqslant c/\log n,$$

where c stands for a positive constant. Hence, applying 10.3 we obtain

$$\begin{aligned} a_{2^r} \leqslant &\left\{2(1+36\rho^{2/(2+\delta)}([2^{(r-1)/5}]))\right\}^{1/(2+\delta)} a_{2^{r-1}} \\ &+18\sigma_{2^{r-1}} + 2\cdot 2^{(r-1)/5}a_1 \\ \leqslant &\, 2^{(r-1)/(2+\delta)}\prod_{j=0}^{r-1}(1+36\rho^{2/(2+\delta)}([2^{i/5}]))^{1/(2+\delta)}a_1 \\ &+c\sigma_1\sum_{j=0}^{r-1}2^{j/2}\prod_{i=j+1}^{r-1}\left\{2(1+9\rho^{2/(2+\delta)}([2^{i/5}]))\right\}^{1/(2+\delta)} \end{aligned}$$

$$+2a_1 \sum_{j=0}^{r-1} 2^{j/5} \prod_{i=j+1}^{r-1} \left\{2(1+9\rho^{2/(2+\delta)}([2^{i/5}]))\right\}^{1/(2+\delta)}$$
$$\leqslant C2^{r/2}\sigma_1 + 2^{r/(2+\delta)} \exp(Cr)^{\delta/(2+\delta)} a_1.$$

This implies the desired inequality.

Remark. If for a certain $q \geqslant 2$, the assumption that $E|X_n|^q < \infty$ is true. Then there exists $C = C(q, \rho(\cdot))$ such that for every $k \geqslant 0$ and $n \geqslant 1$,

$$E \max_{1\leqslant i\leqslant n} |S_k(i)|^q \leqslant Cn^{q/2} \exp\left\{C \sum_{j=0}^{[\log n]} \rho(2^j)\right\} \max_{k<j\leqslant k+n} (E|X_j|^2)^{q/2}$$
$$+n \exp\left\{C \sum_{j=0}^{[\log n]} \rho^{2/q}(2^j)\right\} \max_{k<j\leqslant k+n} E|X_j|^q.$$

See Shao (1995).

10.6 Tail Probability on φ-mixing Sequence

Let $\{X_n, n \geqslant 1\}$ be a φ-mixing sequence, $0 < \eta < 1$. Suppose that there exist an integer p, $1 \leqslant p \leqslant n$, a number $A > 0$ such that

$$\varphi(p) + \max_{p\leqslant j\leqslant n} P\{|S_n - S_j| \geqslant A\} \leqslant \eta.$$

Then, for any $a \geqslant 0$, $b \geqslant 0$, we have

$$P\{\max_{1\leqslant j\leqslant n} |S_j| \geqslant a + A + b\}$$
$$\leqslant \frac{1}{1-\eta} P\{|S_n| \geqslant a\} + \frac{1}{1-\eta} P\left\{\max_{1\leqslant j\leqslant n} |X_j| \geqslant \frac{b}{p-1}\right\},$$

$$P\{|S_n| \geqslant a + A + b\} \leqslant \eta P\{\max_{1\leqslant j\leqslant n} |S_j| \geqslant a\} + P\left\{\max_{1\leqslant j\leqslant n} |X_j| \geqslant \frac{b}{p}\right\}.$$

Proof. Put $E_j = \{\max_{1\leqslant i<j} |S_i| < a + A + b \leqslant |S_j|\}$. Then

$$P\left\{\max_{1\leqslant j\leqslant n} |S_j| \geqslant a+A+b\right\} \leqslant P\{|S_n| \geqslant a\} + \sum_{j=1}^{n-1} P\{E_j \bigcap (|S_n - S_j| \geqslant A+b)\}.$$

Here

$$\begin{aligned}
&\sum_{j=1}^{n-1} P\{E_j \bigcap (|S_n - S_j| \geqslant A + b)\} \\
&\leqslant \sum_{j=1}^{n-p-1} P\{E_j \bigcap (|S_{j+p-1} - S_j| \geqslant b)\} \\
&\quad + \sum_{j=1}^{n-p-1} P\{E_j \bigcap (|S_n - S_{j+p-1}| \geqslant A)\} \\
&\quad + \sum_{j=n-p}^{n-1} P\{E_j \bigcap (|S_n - S_j| \geqslant A + b)\} \\
&\leqslant \sum_{j=1}^{n-1} P\left\{E_j \bigcap (\max_{1\leqslant i\leqslant n} |X_i| \geqslant \frac{b}{p-1})\right\} \\
&\quad + \sum_{j=1}^{n-p-1} P(E_j)(P\{|S_n - S_{j+p-1}| \geqslant A\} + \varphi(p)) \\
&\leqslant P\left\{\max_{1\leqslant i\leqslant n} |X_i| \geqslant \frac{b}{p-1}\right\} + \eta P\left\{\max_{1\leqslant j\leqslant n} |S_j| \geqslant a + A + b\right\}.
\end{aligned}$$

The first inequality is shown.

As for the second one, putting $E'_j = \left\{\max_{1\leqslant i<j} |S_i| < a \leqslant |S_j|\right\}$ and noting that

$$|S_n - S_{j+p+1}| \geqslant ||S_n| - |S_{j-1}| - p \max_{1\leqslant i\leqslant n} |X_i|| \quad \text{for} \quad 1 \leqslant j \leqslant n - p,$$

we have

$$\begin{aligned}
&P\{|S_n| \geqslant a + A + b\} \\
&\leqslant P\left\{|S_n| \geqslant a + A + b,\ \max_{1\leqslant j\leqslant n-p} |S_j| \geqslant a,\ \max_{1\leqslant j\leqslant n} |X_j| \leqslant b/p\right\} \\
&\quad + P\left\{\max_{1\leqslant j\leqslant n} |X_j| > b/p\right\} \\
&\leqslant \sum_{j=1}^{n-p} P\{E'_j \bigcap (|S_n - S_{j+p-1}| > A)\} + P\left\{\max_{1\leqslant j\leqslant n} |X_j| \geqslant b/p\right\} \\
&\leqslant \eta P\left\{\max_{1\leqslant j\leqslant n} |S_j| \geqslant a\right\} + P\left\{\max_{1\leqslant j\leqslant n} |X_j| \geqslant b/p\right\},
\end{aligned}$$

as desired.

See Peligrad (1985).

10.7 Bounds of $2+\delta$-th Moment of Increments of φ-mixing Sequence

Let $\{X_n, n\geqslant 1\}$ be a φ-mixing sequence with $EX_n=0$ and $\sup_n E|X_n|^{2+\delta}<\infty$ for some $\delta>0$. Suppose that

$$\sup_k ES_k^2(n) \leqslant Mn \sup_k EX_k^2 \quad \text{for some} \quad M>0.$$

Then there exists a $C=C(\delta, M, \varphi(\cdot))>0$ such that for each $n\geqslant 1$,

$$\sup_k E|S_k(n)|^{2+\delta} \leqslant Cn^{1+\delta/2} \sup_k E|X_k|^{2+\delta}. \tag{67}$$

Proof. It is easy to see that for $r\geqslant 1$ and $x\geqslant 0$,

$$(1+x)^r \leqslant \sum_{k=0}^{[r]} \binom{r}{k} x^k + \delta_r x^r, \tag{68}$$

where $\delta_r=1$ if r is not an integer, otherwise $\delta_r=0$. We now prove the conclusion by induction on $r=2+\delta$. Assume that the inequality holds for $l\leqslant [r]$, r being a non integer. Denoting $a_m=\sup_k \|S_k(m)\|_r$ and k_0 an integer to be specified later, from (68), we obtain

$$\begin{aligned}
&E|S_k(m)+S_{k+m+k_0}(m)|^r \\
&\leqslant E|S_k(m)|^r + E|S_{k+m+k_0}(m)|^r \\
&\quad + \sum_{j=1}^{[r]} \binom{r}{j} E|S_k(m)|^j |S_{k+m+k_0}(m)|^{r-j} \\
&\leqslant \left(2+2\sum_{j=1}^{[r]} \binom{r}{j} \varphi^{1/r}(k_0)\right) a_m^r \\
&\quad + \sum_{j=1}^{[r]} \binom{r}{j} E|S_k(m)|^j E|S_{k+m+k_0}(m)|^{r-j} \\
&\equiv I_1 + I_2.
\end{aligned} \tag{69}$$

By the induction hypothesis, we have

$$I_2 \leqslant \sum_{j=1}^{[r]} \binom{r}{j} (E|S_k(m)|^{[r]})^{j/[r]} (E|S_{k+m+k_0}(m)|^{[r]})^{(r-j)/[r]}$$
$$\leqslant (m^{[r]/2} \sup_k E|X_k|^{[r]})^{r/[r]} \leqslant cm^{r/2} a_1^r$$

for some $c > 0$. Substituting the above inequality into (69), we obtain

$$a_{2m} \leqslant \left(2 + 2\sum_{j=1}^{[r]} \binom{r}{j} \varphi^{1/r}(k_0) \right)^{1/r} a_m + cm^{1/2} a_1.$$

Now choosing a sufficiently large k_0 and proceeding as in the proof of 10.5, we obtain the desired inequality in this case. Similarly it holds for $[r]+1$. The proof is now complete.

See Shao and Lu (1987).

10.8 Exponential Estimates of Probability on φ-mixing Sequence

Let $\{X_n, n \geqslant 1\}$ be a φ-mixing sequence with $EX_n = 0$ and $|X_n| \leqslant b_n < \infty$. Suppose that there exist $0 < \sigma^2 \leqslant \sigma'^2 < \infty$ such that

$$\sigma^2 n \leqslant \sup_k ES_k(n)^2 \leqslant \sigma'^2 n. \tag{70}$$

Let p, q, k be positive integers with $p = p_n \leqslant n$, $q = q_n = o(p_n)$, $q_n \uparrow \infty$, $k = k_n = [n/(p_n + q_n)]$. Put $b = \max_{1 \leqslant j \leqslant n} b_j$, $\sigma_n^2 = ES_n^2$. Suppose that

$$pb^2 \varphi(q) = o(1), \quad \sum_{j=1}^{k} \varphi^{1/2}(jp) = O(1). \tag{71}$$

Then for $x = x_n$ and small $\varepsilon > 0$ satisfying

$$\frac{4}{\varepsilon} bn\varphi(q) \leqslant x \leqslant \frac{\varepsilon \sigma_n^2}{pb}, \tag{72}$$

$$x^2/n \to \infty \quad \text{as} \quad n \to \infty, \tag{73}$$

we have

$$P\{\max_{1 \leqslant j \leqslant n} |S_j| \geqslant x\} \leqslant 3 \exp\left\{ -\frac{(1-6\varepsilon)x^2}{2\sigma_n^2} \right\}$$

for all large n. If (72) is replaced by

$$x > \varepsilon\sigma_n^2/pb, \tag{74}$$

then

$$P\{\max_{1\leqslant j\leqslant n}|S_j|\geqslant x\}\leqslant 3\exp\left\{-\frac{\varepsilon(1-5\varepsilon)x}{2pb}\right\}.$$

Proof. We always assume that n is large enough. Define

$$\xi_i=\sum_{j=i(p+q)+1}^{(i+1)p+iq}X_j,\quad \eta_i=\sum_{j=(i+1)p+iq+1}^{(i+1)(p+q)}X_j,\quad i=0,1,\cdots,k-1,$$

$$\eta_k=\sum_{j=k(p+q)+1}^{n}X_j.$$

Put σ-algebra $\mathscr{F}_{-1}=\{\varnothing,\Omega\}$, $\mathscr{F}_i=\sigma(X_j,j\leqslant(i+1)p+iq)$, $i=0,1,\cdots,k-1$. Define martingale differences $\gamma_i=\xi_i-E(\xi_i|\mathscr{F}_{i-1})$, $i=0,1,\cdots,k-1$. Write

$$P\{|S_n|\geqslant x\}\leqslant P\left\{\left|\sum_{i=0}^{k-1}\xi_i\right|\geqslant\left(1-\frac{\varepsilon}{2}\right)x\right\}+P\left\{\left|\sum_{i=0}^{k}\eta_i\right|\geqslant\frac{1}{2}\varepsilon x\right\}\equiv J_1+J_2.$$

Consider J_1. We have

$$J_1\leqslant P\left\{\left|\sum_{i=0}^{k-1}\gamma_i\right|\geqslant(1-\varepsilon)x\right\}+P\left\{\left|\sum_{i=0}^{k-1}E(\xi_i|\mathscr{F}_{i-1})\right|\geqslant\frac{1}{2}\varepsilon x\right\}\equiv J_{11}+J_{12}.$$

By 10.1.d (note that $|\xi_i|\leqslant pb$), for any $B_{i-1}\in\mathscr{F}_{i-1}$,

$$|E\xi_iI_{B_{i-1}}|\leqslant 2\varphi(q)pbP(B_{i-1}),$$

which implies

$$|E(\xi_i|\mathscr{F}_{i-1})|\leqslant 2\varphi(q)pb\quad\text{a.s.},\quad i=0,1,\cdots,k-1. \tag{75}$$

Using condition (72), we obtain $J_{12}=0$.

We now estimate J_{11}. It is easy to verify that for $0<\lambda\leqslant((1+\varepsilon)pb)^{-1}$ and large n,

$$\zeta_j\equiv\exp\left(\lambda\sum_{i=0}^{j}\gamma_i\right)\exp\left\{-\frac{\lambda^2}{2}\left(1+\frac{1}{2}(1+\varepsilon)pb\lambda\right)\sum_{i=0}^{j}E(\gamma_i^2|\mathscr{F}_{i-1})\right\}$$

where

$$j = 0, 1, \cdots, k-1,$$

possesses the supermartingale property by noting that $|\gamma_i| \leqslant (1+\varepsilon)pb$ a.s. and expanding $E\{\exp(\lambda\gamma_i)|\mathscr{F}_{i-1}\}$. Write

$$\begin{aligned} & P\left\{\sum_{i=0}^{k-1} \gamma_i \geqslant (1-\varepsilon)x\right\} \\ = & P\Bigg\{\zeta_{k-1} \geqslant \exp(\lambda(1-\varepsilon)x) \\ & \cdot \exp\left\{-\frac{\lambda^2}{2}\left(1+\frac{1}{2}(1+\varepsilon)pb\lambda\right)\sum_{i=0}^{k-1} E(\gamma_i^2|\mathscr{F}_{i-1})\right\}\Bigg\}. \end{aligned} \tag{76}$$

We have $|E(\xi_i^2|\mathscr{F}_{i-1}) - E\xi_i^2| \leqslant 2\varphi(q)(pb)^2$ a.s. by a proof similar to (75). Hence using conditions (70) and (71), we obtain

$$\sum_{i=0}^{k-1} E(\xi_i^2|\mathscr{F}_{i-1}) = (1+o(1))\sum_{i=0}^{k-1} E\xi_i^2 \qquad \text{a.s.}$$

Moreover (75) implies that

$$\sum_{i=0}^{k-1} (E(\xi_i|\mathscr{F}_{i-1}))^2 \leqslant 4\varphi(q)^2 p^2 b^2 n = o(n) \quad \text{a.s.}$$

Thus

$$\begin{aligned} \sum_{i=0}^{k-1} E(\gamma_i^2|\mathscr{F}_{i-1}) &= \sum_{i=0}^{k-1} E(\xi_i^2|\mathscr{F}_{i-1}) - \sum_{i=0}^{k-1} (E(\xi_i|\mathscr{F}_{i-1}))^2 \\ &= (1+o(1))\sum_{i=0}^{k-1} E\xi_i^2 \quad \text{a.s.} \end{aligned}$$

We have

$$\sum_{i=0}^{k-1} E\xi_i^2 = E\left(\sum_{i=0}^{k-1} \xi_i\right)^2 - 2\sum_{0\leqslant i<j\leqslant k-1} E\xi_i\xi_j,$$

$$\begin{aligned} \sum_{0\leqslant i<j\leqslant k-1} E\xi_i\xi_j &\leqslant \sum_{i=0}^{k-2}\left(|E\xi_i\xi_{i+1}| + \sum_{j=i+2}^{k-1} |E\xi_i\xi_j|\right) \\ &\leqslant 2\sigma' p^{\frac{1}{2}} pb \sum_{i=0}^{k-2}(\varphi(q) + \sum_{j=i+2}^{k-1} \varphi((j-i-1)p)) \\ &= o(n) \end{aligned}$$

by 10.1.d and (71). Furthermore

$$E\left(\sum_{i=0}^{k-1}\xi_i\right)^2 = ES_n^2 - 2ES_n\left(\sum_{i=0}^{k}\eta_i\right) + E\left(\sum_{i=0}^{k}\eta_i\right)^2.$$

By condition (70) and the second equality in (71), we have

$$E\left(\sum_{i=0}^{k}\eta_i\right)^2 \leqslant 2E\left(\sum_{i=0}^{k-1}\eta_i\right)^2 + 2E\eta_k^2 = O(kq+p).$$

In fact $E\eta_k^2 = O(p)$ and

$$\begin{aligned}E\left(\sum_{i=0}^{k-1}\eta_i\right)^2 &= \sum_{i=0}^{k-1}E\eta_i^2 + 2\sum_{0\leqslant i<j\leqslant k-1}E\eta_i\eta_j\\ &\leqslant \sigma'^2 kq + 2\sigma'^2 q\sum_{j=1}^{k-1}(k-j)\varphi^{\frac{1}{2}}(jp)\\ &= O(kq).\end{aligned}$$

Hence

$$E\left(\sum_{i=0}^{k-1}\xi_i\right)^2 = (1+o(1))\sigma_n^2.$$

Combining these estimates yields

$$\sum_{i=0}^{k-1}E(\gamma_i^2|\mathscr{F}_{i-1}) = (1+o(1))\sigma_n^2 \qquad \text{a.s.} \tag{77}$$

Inserting it into (76) and noting that $P\{\zeta_{k-1}\geqslant\alpha\}\leqslant\alpha^{-1}$ for any $\alpha>0$, we obtain

$$\begin{aligned}&P\left\{\sum_{i=0}^{k-1}\gamma_i \geqslant (1-\varepsilon)x\right\}\\ &\leqslant \exp\left\{-\lambda(1-\varepsilon)x + \frac{\lambda^2}{2}\left(1+\frac{1}{2}(1+\varepsilon)pb\lambda\right)(1+\varepsilon)\sigma_n^2\right\}.\end{aligned} \tag{78}$$

Choosing $\lambda = x/((1+\varepsilon)\sigma_n^2)$, we have $\lambda\leqslant\varepsilon((1+\varepsilon)pb)^{-1}$ by (72). Then

$$P\left\{\sum_{i=0}^{k-1}\gamma_i \geqslant (1-\varepsilon)x\right\} \leqslant \exp\left\{-\frac{(1-4\varepsilon)x^2}{2\sigma_n^2}\right\}.$$

Replacing X_j with $-X_j$, we obtain

$$J_{11} = P\left\{\left|\sum_{i=0}^{k-1}\gamma_i\right| \geqslant (1-\varepsilon)x\right\} \leqslant 2\exp\left\{-\frac{(1-4\varepsilon)x^2}{2\sigma_n^2}\right\}.$$

This also is a bound for J_1 by recalling that $J_{12} = 0$.

For J_2, it is clear that $3^{-1}\exp\{-(1-4\varepsilon)x^2/(2\sigma_n^2)\}$ is one of its upper bounds in view of $q_n = o(p_n)$. Thus we have proved

$$P\{|S_n| \geqslant x\} \leqslant \frac{7}{3}\exp\left\{-\frac{(1-4\varepsilon)x^2}{2\sigma_n^2}\right\}. \tag{79}$$

In order to get the first desired inequality from (79), we use 10.6. Choose an integer p_0 such that $\varphi(p_0) \leqslant 10^{-1}$. Using conditions (70) and (73), we obtain

$$\max_{1\leqslant i\leqslant n} P\left\{|S_n - S_i| \geqslant \frac{\varepsilon x}{2(1+\varepsilon)}\right\} \leqslant \frac{4(1+\varepsilon)^2\sigma'^2 n}{\varepsilon^2 x^2} \leqslant \frac{1}{10}$$

for large n. So η in 10.6 can be chosen as 5^{-1}. Furthermore

$$\max_{1\leqslant i\leqslant n}|X_i| \leqslant b \leqslant \frac{\varepsilon\sigma'^2 n}{px} \leqslant \frac{\varepsilon\sigma'^2 k}{x^2}x = o(x) \tag{80}$$

by (72) and (73). Hence for large n,

$$P\left\{\max_{1\leqslant i\leqslant n}|X_i| \geqslant \frac{\varepsilon x}{2(1+\varepsilon)(p_0-1)}\right\} = 0.$$

By 10.6 we obtain the first desired inequality:

$$\begin{aligned}
&P\{\max_{1\leqslant i\leqslant n}|S_i| \geqslant x\}\\
&\leqslant \frac{5}{4}P\left\{|S_n| \geqslant \frac{x}{1+\varepsilon}\right\} + \frac{5}{4}P\left\{\max_{1\leqslant i\leqslant n}|X_i| \geqslant \frac{\varepsilon x}{2(1+\varepsilon)(p_0-1)}\right\}\\
&\leqslant 3\exp\left\{-\frac{(1-6\varepsilon)}{2\sigma_n^2}x^2\right\}.
\end{aligned}$$

We will prove the second inequality next. Clearly we can assume that $x/pb > \delta$ for some $\delta > 0$. If (74) is satisfied, choosing $\lambda = \varepsilon((1+\varepsilon)pb)^{-1}$ in (78), we have

$$P\left\{\sum_{i=0}^{k-1}\gamma_i \geqslant (1-\varepsilon)x\right\} \leqslant \exp\left\{-\frac{\varepsilon(1-4\varepsilon)x^2}{2pb}\right\}. \tag{81}$$

By imitating the preceding procedure and replacing (80) by $\max_{1\leqslant i\leqslant n} |X_i| \leqslant b < x/pb = o(x)$, we obtain (77) from (81). The inequality has now been proven.

See Lin (1991).

References

Doukhan P. 1994. Mixing: Properties and Examples. Lecture Notes in Statist., 85. Springer-Verlag.

Lin Z Y. 1982. Weak convergence of weak dependent sequences with possibly unbounded variances. Chin, Ann. Math. 3A.

Lin Z Y. 1991. On the increments of partial sums of a φ-mixing sequence. Probab. Theory Appl., 36: 344-354.

Lin Z Y, Lu C L. 1996. Limit Theory for Mixing Dependent random variables. Dordrecht & Beijing: Kluwer Academic Publishers & Science Press.

Peligrad M. 1985. An invariance principle for φ-mixing sequences. Ann. Probab., 13: 1304-1313.

Peligrad M. 1987. On the central limit theorem for ρ-mixing sequences of random variables. Ann. Probab., 15: 1387-1394.

Shao Q M. 1995. Maximal inequality for partial sums of ρ-mixing sequences. Ann. Probab., 23: 948-965.

Shao Q M, Lu C R. 1987. Strong approximations for partial sums of weakly dependent random variables. Scientia Sinica, 30: 575-587.

Chapter 11

Inequalities Related to Associative Variables

In this chapter, we introduce another class of dependent variables. Two r.v.'s X and Y are said to be positive quadrant dependent (PQD) if $P(X > x, Y > y) \geqslant P(X > x)P(Y > y)$ for any x, y; and negative quadrant dependent (NQD) if $P(X > x, Y > y) \leqslant P(X > x)P(Y > y)$.

A set of n r.v.'s $X_1,\cdots,X_n$ is said to be (positive) associated (PA) if for any coordinatewise nondecreasing functions f and g on R^n, $\mathrm{Cov}(f(X_1, \cdots, X_n), g(X_1, \cdots, X_n)) \geqslant 0$, whenever the covariance exists. The set is said to be negative associated (NA) if for any disjoint A, $B \subset \{1, \cdots, n\}$ and any nondecreasing functions f on R^A and g on R^B, $\mathrm{Cov}(f(X_k, k \in A), g(X_j, j \in B)) \leqslant 0$.

An infinite family of r.v.'s is said to be linearly positive quadrant dependent (LPQD) if for any disjoint integer sets A, B and positive a_j's, $\sum\limits_{k\in A} a_k X_k$ and $\sum\limits_{j\in B} a_j X_j$ are PQD; linearly negative quadrant dependent (LNQD) is obviously in an analogous manner. An infinite family of r.v.'s is said to be (positive) associated (resp. negative associated) if every finite subfamily is PA (resp. NA).

Clearly, for a pair of r.v.'s PQD (resp. NQD) is equivalent to PA (resp. NA). For a family of r.v.'s, PA (resp. NA) implies LPQD (resp. LNQD).

11.1 Covariance of PQD Varaibles

If X and Y are PQD (resp. NQD) r.v.'s, then

$$EXY \geqslant EXEY \qquad (\text{resp.} \quad EXY \leqslant EXEY),$$

whenever the expectations exist. The equality holds if and only if X and Y are independent.

Proof. Consider only the PQD case. If F denotes the joint and F_X and F_Y the marginal distributions of X and Y, then we have

$$EXY - EXEY = \iint (F(x,y) - F_X(x)F_Y(y))\mathrm{d}x\mathrm{d}y, \tag{82}$$

which implies the desired inequality immediately from the definition of PQD.

Now suppose that the equality holds. Then $F(x,y) = F_X(x)F_Y(y)$ except possibly on a set of Lebesgue measure zero. From the fact that d.f.'s are right continuous, it is easily seen that if two d.f.'s agree almost everywhere with respect to the Lebesgue measure, they must agree everywhere. Thus X and Y are independent.

See Lehmann (1966).

11.2 Probability of Quadrant on PA (NA) Sequence

Let $X_1,\cdots,X_n$ be PA (resp. NA) r.v.'s, $Y_j = f_j(X_1,\cdots,X_n)$ and f_j be nondecreasing, $j = 1, \cdots, k$. Then for any $x_1, \cdots, x_k$,

$$P\left\{\bigcap_{j=1}^{k}(Y_j \leqslant x_j)\right\} \geqslant \prod_{j=1}^{k} P\{Y_j \leqslant x_j\}$$

$$\left(\text{resp.}\quad P\left\{\bigcap_{j=1}^{k}(Y_j \leqslant x_j)\right\} \leqslant \prod_{j=1}^{k} P\{Y_j \leqslant x_j\}\right),$$

$$P\left\{\bigcap_{j=1}^{k}(Y_j > x_j)\right\} \geqslant \prod_{j=1}^{k} P\{Y_j > x_j\}$$

$$\left(\text{resp.}\quad P\left\{\bigcap_{j=1}^{k}(Y_j > x_j)\right\} \leqslant \prod_{j=1}^{k} P\{Y_j > x_j\}\right).$$

Proof. Consider only the PA case. Clearly, $Y_1,\cdots,Y_k$ are PA. Let $A_j = \{Y_j > x_j\}$. Then $I_j = I(A_j)$ is nondecreasing in Y_j, and so, $I_1,\cdots,I_k$ are PA. Investigate increasing functions $f(t_1,\cdots,t_k)= t_1$ and $g(t_1,\cdots,t_k)= t_2\cdots t_k$. $f(I_1,\cdots,I_k)$ and $g(I_1,\cdots,I_k)$ are PA and hence by 11.1,

$$E(I_1 I_2 \cdots I_k) \geqslant E(I_1)E(I_2\cdots I_k).$$

Repeating use of this argument yields $E(I_1 I_2 \cdots I_k) \geqslant E(I_1) \cdots E(I_k)$. The second desired inequality is proven.

As for the first one, consider $1 - I_j$ instead of I_j. Let $\bar{f}(t_1,\cdots,t_k) = 1 - f(1-t_1,\cdots,1-t_k)$, $\bar{g}(t_1,\cdots,t_k) = 1 - g(1-t_1,\cdots,1-t_k)$, which are both increasing. Then

$$\begin{aligned}&\mathrm{Cov}(f(1-I_1,\cdots,1-I_k), g(1-I_1,\cdots,1-I_k))\\ =&\mathrm{Cov}(\bar{f}(I_1,\cdots,I_k), \bar{g}(I_1,\cdots,I_k)) \geqslant 0,\end{aligned}$$

which implies the first desired inequality.

See Esary et al. (1967).

11.3 Estimates of c.f.'s on LPQD (LNQD) Sequence

Let $X_1, \cdots, X_n$ be LPQD or LNQD r.v.'s, $\varphi_j(t_j)$ and $\varphi(t_1, \cdots, t_n)$ be the c.f.'s of X_j and $(X_1, \cdots, X_n)$ respectively. Then

$$|\varphi(t_1,\cdots,t_n) - \prod_{j=1}^{n} \varphi_j(t_j)| \leqslant \sum_{1 \leqslant k < l \leqslant n} |t_k t_l \mathrm{Cov}(X_k, X_l)|.$$

Proof. We show the inequality by induction on n. For $n = 2$, integration by parts, analogously to (82) in 11.1, yields

$$\mathrm{Cov}(\mathrm{e}^{\mathrm{i}t_1 X_1}, \mathrm{e}^{\mathrm{i}t_2 X_2}) = \iint \mathrm{i}t_1 \mathrm{e}^{\mathrm{i}t_1 x_1} \mathrm{i}t_2 \mathrm{e}^{\mathrm{i}t_2 x_2} H(x_1, x_2) \mathrm{d}x_1 \mathrm{d}x_2,$$

where $H(x_1, x_2) = P(X_1 > x_1, X_2 > x_2) - P(X_1 > x_1)P(X_2 > x_2)$. The triangle inequality, the pointwise positivity of H and (82) in 11.1 yield the desired inequality for $n = 2$.

Choose a nontrivial subset A of $\{1, \cdots, n\}$ so that the t_j's have a common sign for $j \in A$ and a common sign for $j \in \bar{A} = \{1, \cdots, n\} - A$. Without loss of generality assume that $A = \{1, \cdots, m\}$ $(1 \leqslant m < n)$ (by relabeling indices if necessary). Define $Y_1 = \sum\limits_{j=1}^{m} t_j X_j$, $Y_2 = \sum\limits_{j=m+1}^{n} t_j X_j$ and $\varphi_Y = E\exp(\mathrm{i}Y)$. Then

$$
\begin{aligned}
&|\varphi(t_1,\cdots,t_n)-\prod_{j=1}^{n}\varphi_j(t_j)| \\
&\leqslant |\varphi(t_1,\cdots,t_n)-\varphi_{Y_1}\varphi_{Y_2}|+|\varphi_{Y_1}|\left|\varphi_{Y_2}-\prod_{j=m+1}^{n}\varphi_j(t_j)\right| \\
&\quad+\left|\prod_{j=m+1}^{n}\varphi_j(t_j)||\varphi_{Y_1}-\prod_{j=1}^{m}\varphi_j(t_j)\right| \\
&\leqslant |\mathrm{Cov}(Y_1,Y_2)|+\sum_{m+1\leqslant k<l\leqslant n}|t_kt_l\mathrm{Cov}(X_k,X_l)| \\
&\quad+\sum_{1\leqslant k<l\leqslant m}|t_kt_l\mathrm{Cov}(X_k,X_l)| \\
&\leqslant \sum_{1\leqslant k<l\leqslant n}|t_kt_l\mathrm{Cov}(X_k,X_l)|.
\end{aligned}
$$

See Newman (1984).

11.4 Maximal Partial Sums of PA Sequence

Let $X_1,\cdots,X_n$ be PA r.v.'s with means zero and finite variances. Put $S_k=\sum_{j=1}^{k}X_j$.

11.4.a. Put $M_n=\max_{1\leqslant k\leqslant n}S_k$. We have

$$EM_n^2\leqslant \mathrm{Var}S_n.$$

Proof. Define $K_n=\min\{0,X_2+\cdots+X_n,X_3+\cdots+X_n,\cdots,X_n\}$, $L_n=\max\{X_2,X_2+X_3,\cdots,X_2+\cdots+X_n\}$, $J_n=\max\{0,L_n\}$, and note that $K_n=X_2+\cdots+X_n-J_n$ is a nondecreasing function of X_j's so that $\mathrm{Cov}(X_1,K_n)\geqslant 0$; thus

$$
\begin{aligned}
EM_n^2&=E(X_1+J_n)^2=\mathrm{Var}X_1+2\mathrm{Cov}(X_1,J_n)+EJ_n^2\\
&=\mathrm{Var}X_1+2\mathrm{Cov}(X_1,X_2+\cdots+X_n)-2\mathrm{Cov}(X_1,K_n)+EJ_n^2\\
&\leqslant \mathrm{Var}X_1+2\mathrm{Cov}(X_1,X_2+\cdots+X_n)+EL_n^2. \qquad (83)
\end{aligned}
$$

The proof is completed by induction on n since the induction hypothesis implies that $EL_n^2\leqslant \mathrm{Var}(X_2+\cdots+X_n)$, which together with (83) yields the desired inequality.

11.4.b. Put $s_n^2 = ES_n^2$. We have for any $x \geqslant \sqrt{2}$,

$$P\{\max_{1\leqslant j\leqslant n} |S_j| \geqslant xs_n\} \leqslant 2P\{|S_n| \geqslant (x-\sqrt{2})s_n\}.$$

Proof. Put $S_n^* = \max(0, S_1, \cdots, S_n)$ and note that for $0 \leqslant x_1 < x_2$,

$$\begin{aligned} P\{S_n^* \geqslant x_2\} &\leqslant P\{S_n \geqslant x_1\} + P\{S_{n-1}^* \geqslant x_2, S_{n-1}^* - S_n > x_2 - x_1\} \\ &\leqslant P\{S_n \geqslant x_1\} + P\{S_{n-1}^* \geqslant x_2\}P\{S_{n-1}^* - S_n > x_2 - x_1\} \\ &\leqslant P\{S_n \geqslant x_1\} + P\{S_n^* \geqslant x_2\}E(S_{n-1}^* - S_n)^2/(x_2-x_1)^2. \end{aligned}$$

Here we have used the fact that S_{n-1}^* and $S_n - S_{n-1}^*$ are PA since they are both nondecreasing functions of the X_j's. Now 11.4.a with X_j replaced by $Y_j = -X_{n-j+1}$ yields

$$E(S_{n-1}^* - S_n)^2 = E(\max(Y_1, Y_1+Y_2, \cdots, Y_1+\cdots+Y_n)^2) \leqslant ES_n^2$$

and thus we have, for $(x_2-x_1)^2 \geqslant s_n^2$,

$$P\{S_n^* \geqslant x_2\} \leqslant (1 - s_n^2/(x_2-x_1)^2)^{-1} P\{S_n \geqslant x_1\}. \qquad (84)$$

By adding to (84) the analogous inequality with each X_j replaced by $-X_j$, and by choosing $x_2 = xs_n$, $x_1 = (x-\sqrt{2})s_n$, we obtain the desired inequality.

See Newman and Wright (1981).

11.5 Variance of Increment of LPQD Sequence

Let $\{X_n, n \geqslant 1\}$ be a sequence of LPQD r.v.'s with $EX_n = 0$. Put $S_k(n) = \sum_{j=k+1}^{k+n} X_j$ and $\mu(n) = \sup_{k\geqslant 1} \sum_{j:|j-k|\geqslant n} \operatorname{Cov}(X_j, X_k)$.

$$\sup_{k\geqslant 1} ES_k(n)^2 \leqslant 4n \left(\sup_{n\geqslant 1} EX_n^2 + \sum_{i=1}^{[\log n]} \max_{(n/2^i)^{1/3}\leqslant j\leqslant n/2^{i-1}} \mu(j) \right).$$

Proof. Put $\| X \|_p = (E|X|^p)^{1/p}$ $(p > 0)$, $\tau_m = \sup_{k\geqslant 1} \| S_k(m) \|_2$ and $m_1 = m + [m^{1/3}]$. Write

$$S_k(2m) = S_k(m) + S_{k+m}([m^{1/3}]) + S_{k+m_1}(m) - S_{k+2m}([m^{1/3}]).$$

We have

$$\|S_k(2m)\|_2 \leqslant \|S_k(m) + S_{k+m_1}(m)\|_2 + 2[m^{1/3}]\tau_1$$

and

$$\begin{aligned}E(S_k(m)+S_{k+m_1}(m))^2 &\leqslant 2\tau_m^2+2ES_k(m)S_{k+m_1}(m)\\&\leqslant 2\tau_m^2+2\sum_{j=[m^{1/3}]+1}^{m_1}\mu(j).\end{aligned}$$

Recurrently, for each integer $r>0$, we obtain

$$\begin{aligned}\tau_{2^r}^2 &\leqslant 2\left\{2^r\tau_1^2+\sum_{i=1}^{r}2^i\sum_{j=[2^{(r-i)/3}]+1}^{2^{r-i+1}}\mu(j)+4\sum_{i=1}^{r}2^{2(i-1)/3}\tau_1^2\right\}\\&\leqslant 4\cdot 2^r\left(\tau_1^2+\sum_{i=1}^{r}\max_{2^{(r-i)/3}\leqslant j\leqslant 2^{r-i+1}}\mu(j)\right),\end{aligned}$$

which implies the desired estimate.

See Lin (1996).

11.6 Expectation of Convex Function of Sum of NA Sequence

Let $X_1,\cdots,X_n$ be NA r.v.'s, and $X_1^*,\cdots,X_n^*$ be independent r.v.'s. For each $i=1,\cdots,n$, X_i^* and X_i have identical distribution. Then for any convex function f in R^1,

$$Ef\left(\sum_{i=1}^{n}X_i\right)\leqslant Ef\left(\sum_{i=1}^{n}X_i^*\right), \tag{85}$$

if the expectations exist in the formula above.

If f is nondecreasing, then

$$Ef\left(\max_{1\leqslant k\leqslant n}\sum_{i=1}^{k}X_i\right)\leqslant Ef\left(\max_{1\leqslant k\leqslant n}\sum_{i=1}^{k}X_i^*\right), \tag{86}$$

if the expectations indicated exist.

Proof. We just prove (85) and for the proof of (86) one can refer to (Shao, 2000). By induction, we shall now prove

$$Ef(X_1+X_2)\leqslant Ef(X_1^*+X_2^*). \tag{87}$$

Let (Y_1, Y_2) be an independent copy of (X_1, X_2). Then we have

$$\begin{aligned}
&f(X_1+X_2)+f(Y_1+Y_2)-f(X_1+Y_2)-f(Y_1+X_2)\\
=&\int_{X_2}^{Y_2}(f_+'(Y_1+t)-f_+'(X_1+t))\,dt\\
=&\int_{-\infty}^{\infty}(f_+'(Y_1+t)-f_+'(X_1+t))(I(Y_2>t)-I(X_2>t))\,dt,
\end{aligned}$$

in which $f_+'(x)$ is the right derivative of $f(x)$. It is nondecreasing. By positive association and the Fubini theorem, we obtain

$$\begin{aligned}
&2(Ef(X_1+X_2)-Ef(X_1^*+X_2^*))\\
=&E(f(X_1+X_2)+f(Y_1+Y_2)-f(X_1+Y_2)-f(Y_1+X_2))\\
=&2\int_{-\infty}^{\infty}Cov(f_+'(X_1+t),I(X_2\geqslant t))\,dt\leqslant 0.
\end{aligned}$$

Thus (87) has been proved.

Remark. We can obtain a series of important inequalities about NA r.v.'s as consequences of the result above. For example, for $1<p\leqslant 2$,

$$E\max_{1\leqslant k\leqslant n}\left|\sum_{i=1}^{k}X_i\right|^p\leqslant 2E\left|\sum_{i=1}^{n}X_i^*\right|^p,$$

$$E\max_{1\leqslant k\leqslant n}\left|\sum_{i=1}^{k}X_i\right|^p\leqslant 2^{3-p}\sum_{i=1}^{n}E|X_i^*|^p;$$

for $p>2$,

$$E\max_{1\leqslant k\leqslant n}\left|\sum_{i=1}^{k}X_i\right|^p\leqslant 2(15p/\log p)^p\left\{\left(\sum_{i=1}^{n}EX_i^2\right)^{p/2}+\sum_{i=1}^{n}E|X_i|^p\right\}.$$

Besides, put $S_n=\sum\limits_{i=1}^{n}X_i$, $B_n=\sum\limits_{i=1}^{n}EX_i^2$. Then for any $x>0$, $a>0$ and $0<\alpha<1$,

$$P\left\{\max_{1\leqslant k\leqslant n}S_k\geqslant x\right\}\leqslant P\left\{\max_{1\leqslant k\leqslant n}X_k>a\right\}+\frac{1}{1-\alpha}$$

$$\exp\left\{-\frac{x^2\alpha}{2(ax+B_n)}\left(1+\frac{2}{3}\log\left(1+\frac{ax}{B_n}\right)\right)\right\},$$

$$P\left\{\max_{1\leqslant k\leqslant n}|S_k|\geqslant x\right\}\leqslant 2P\left\{\max_{1\leqslant k\leqslant n}|X_k|>a\right\}+\frac{2}{1-\alpha}$$
$$\exp\left\{-\frac{x^2\alpha}{2(ax+B_n)}\left(1+\frac{2}{3}\log\left(1+\frac{ax}{B_n}\right)\right)\right\}.$$

See Shao (2000).

11.7 Marcinkiewicz-Zygmund-Burkholder Inequality for NA Sequence

Let $X_1,\cdots,X_n$ be zero-mean-value NA r.v.'s. Then for $r\geqslant 1$,

$$E\left|\sum_{i=1}^n X_i\right|^r\leqslant A_r E\left(\sum_{i=1}^n X_i^2\right)^{r/2}.$$

Proof. Let X,Y be zero-mean-value NA r.v.'s, and $f(x)$ be a convex function. Similar to the proof in 11.6, we have

$$Ef(X+Y)\leqslant Ef(X-Y).$$

Now let $\varepsilon_j, j=1,\cdots,n$ be i.i.d. r.v.'s, of which the distribution is $P(\varepsilon_1=1)=P(\varepsilon_1=-1)=\frac{1}{2}$. Assume that $\{\varepsilon_j\}$ and $\{X_j\}$ are independent. Take

$$X=\sum_{j=1}^n X_j I(\varepsilon_j=1),\quad Y=\sum_{j=1}^n X_j I(\varepsilon_j=-1).$$

Then

$$X+Y=\sum_{j=1}^n X_j,\quad X-Y=\sum_{j=1}^n \varepsilon_j X_j.$$

Under the condition of $\{\varepsilon_j\}$, X and Y are NA r.v.'s. Therefore

$$\begin{aligned}Ef\left(\sum_{j=1}^n X_j\right)&=Ef(X+Y)=E(E(f(X+Y)|\varepsilon_1,\varepsilon_2,\cdots))\\&\leqslant E(E(f(X-Y)|\varepsilon_1,\varepsilon_2,\cdots))\\&=Ef(X-Y)=Ef\left(\sum_{j=1}^n\varepsilon_j X_j\right).\end{aligned}$$

Choose $f(x) = |x|^r$. By the Khintchine inequality 9.6 we obtain

$$E\left|\sum_{i=1}^{n} X_i\right|^r \leqslant E\left|\sum_{j=1}^{n} \varepsilon_j X_j\right|^r \leqslant A_r E\left(\sum_{i=1}^{n} X_i^2\right)^{r/2}.$$

See Zhang (2000).

References

Birkel T. 1988. Momemt bounds for associated sequences. Ann. Probab., 16: 1184-1193.

Esary J, Proschan F, Walkup D. 1967. Association of random variables with application. Ann. Math. Statist., 38: 1466-1474.

Lehmann E L. 1966. Some concepts of dependence. Ann. Math. Statist., 37: 1137-1153.

Lin Z Y. 1996. An invariance principle for positive dependent Random Variables. Chin. Ann. Math., 17A: 487-494.

Matula M. 1992. A note on the almot sure convergence of sums of negatively dependent random variables. Statist. & Probab. Let., 15: 209-213.

Newman C M. 1984. Asymptotic independence and limit theorems for positively and negatively dependent random variables//Inequalities in Statistics and Probability. Tong Y L ed., 127-140.

Newman C M, Wright A L. 1981. An invariance principle for certain dependent sequences. Ann. Probab., 9: 671-675.

Shao Q M. 2000. A comparison theorem on moment inequalities between negative associated and independent random variables. J. Theor. Probab., 13: 343-356.

Su C, Zhao L C, Wang Y B. 1997. Moment inequalities and weak convergence for NA sequence. Science in China, 40(A): 347-355.

Zhang L X. 2000. A functional central limit theorem for asymptotically negatively dependent random variables. Acta. Math. Hungar., 86: 237-259.

Chapter 12

Inequalities about Stochastic Processes and Banach Space Valued Random Variables

In this chapter, we introduce some important inequalities about stochastic processes and Banach space valued random elements. Because the image space of the random elements is more complicated than those for random variables or vectors, the proofs are generally more involved or need specific approaches. Therefore, proofs of some inequalities will be omitted and some references are given.

Herewith, we first recall the definitions of Wiener process and Poison Process below.

A stochastic process $\{W(t), t \geqslant 0\}$ defined on a probability space $(\Omega, \mathscr{F}, P)$ is called a Wiener process (or a Brownian motion) if

(i) for any $\omega \in \Omega$, $W(0, \omega) \equiv 0$ and for any $0 \leqslant s \leqslant t$, $W(t) - W(s) \in N(0, t-s)$;

(ii) sample functions $W(t, \omega)$ are continuous on $[0, \infty)$ with probability one;

(iii) for $0 \leqslant t_1 < t_2 \leqslant t_3 < t_4 \leqslant \cdots \leqslant t_{2n-1} < t_{2n}$, the increments $W(t_2) - W(t_1)$, $W(t_4) - W(t_3)$, $\cdots$, $W(t_{2n}) - W(t_{2n-1})$ are independent.

A stochastic process $\{N(t), t \geqslant 0\}$ is called a Poisson process if for all $k \geqslant 1$ and all $0 \leqslant t_1 \leqslant t_2 \leqslant \cdots \leqslant t_k$, $N(t_1)$, $N(t_2) - N(t_1)$, $\cdots$, $N(t_k) - N(t_{k-1})$ are independent Poisson r.v.'s with means t_1, $t_2 - t_1$, $\cdots$, $t_k - t_{k-1}$ respectively.

12.1 Probability Estimates of Supremums of a Wiener Process

12.1.a. For any $x \geqslant 0$,

$$P\{\sup_{0 \leqslant s \leqslant t} W(s) \geqslant x\} = 2P\{W(t) \geqslant x\},$$

$$P\{\inf_{0\leqslant s\leqslant t} W(s) \leqslant -x\} = 2P\{W(t) \leqslant -x\},$$

$$P\{\sup_{0\leqslant s\leqslant t} |W(s)| \geqslant x\} \leqslant 2P\{|W(t)| \geqslant x\} = 4P\{W(1) \geqslant xt^{-1/2}\}.$$

Proof. We shall only prove the first identity, since the second is a dual case by symmetry of a Wiener Process and the third one (inequality) follows by combining the two identities. Also, when $x = 0$, the two equalities and the inequality are trivially true. Thus, we assume $x > 0$ in the proof.

We first prove **Reflection Principle of Wiener Process**. If $W(s)$, $0 \leqslant s \leqslant t \leqslant \infty$ is a Wiener process, x is a real number and S is a stopping time, then $W'(s)$ is also a Wiener process on $[0, t]$, where

$$W'(s) = \begin{cases} W(s), & \text{if } 0 \leqslant s \leqslant S \wedge t, \\ 2W(S) - W(s), & \text{if } S \wedge t < s \leqslant t. \end{cases}$$

Especially, when $W(S) = x$, $W'(s)$ is the process obtained by reflecting $W(s)$, $s > S$ about the straight line x.

By strong Markovian property of a Wiener process, $W(s)$, $s > S \wedge t$ is independent of $W(s)$, $s \leqslant S \wedge t$. By symmetry of a Wiener process, $W(S+s) - W(S)$, $s \geqslant 0$ has the same distribution as $W(S) - W(S+s)$, $s \geqslant 0$. Thus, W and W' are identically distributed.

Now, let's return to our proof of the first identity. Obviously, we have

$$P\Big\{\sup_{0\leqslant s\leqslant t} W(s) \geqslant x\Big\} = P\Big\{\sup_{0\leqslant s\leqslant t} W(s) \geqslant x; W(t) \geqslant x\Big\} + P\Big\{\sup_{0\leqslant s\leqslant t} W(s) \geqslant x,\ W(t) < x\Big\}.$$

Since $\{W(t) \geqslant x\}$ implies $\{\sup_{0\leqslant s\leqslant t} W(s) \geqslant x\}$, we have

$$P\Big\{\sup_{0\leqslant s\leqslant t} W(s) \geqslant x;\ W(t) \geqslant x\Big\} = P(W(t) \geqslant x).$$

By the reflection principle, we have

$$P\Big\{\sup_{0\leqslant s\leqslant t} W(s) \geqslant x, W(t) < x\Big\} = P\Big\{\sup_{0\leqslant s\leqslant t} W'(s) \geqslant x, W'(t) > x\Big\} = P(W(t) > x).$$

The proof is complete.

12.1.b. For any $x > 0$,

$$P\{\sup_{0\leqslant t\leqslant 1} W(t) \leqslant x\} = \frac{2}{\sqrt{2\pi}}\int_0^x e^{-u^2/2}du;$$

$$\frac{4}{\pi}\left(e^{-\pi^2/8x^2} - \frac{1}{3}e^{-9\pi^2/8x^2}\right) \leqslant P\{\sup_{0\leqslant t\leqslant 1}|W(t)| \leqslant x\} \leqslant \frac{4}{\pi}e^{-\pi^2/8x^2}.$$

Proof. The first equality follows from the first equality of 12.1.a. To prove the inequalities, we first prove that

$$P\Big\{\sup_{0\leqslant t\leqslant 1}|W(t)| < x\Big\} = \frac{1}{\sqrt{2\pi}}\sum_{k=-\infty}^{\infty}(-1)^k\int_{(2k-1)x}^{(2k+1)x} e^{-t^2/2}dt.$$

Define $\tau_0 = \inf\{t \leqslant 1; |W(t)| > x\}$ and $\tau_1 = \inf\{t \leqslant 1; W(t) < -x\}$. By induction, define $\zeta_k = \inf\{t \in (\tau_k, 1];\ W(t) > x\}$ and $\tau_{k+1} = \inf\{t \in (\zeta_k, 1];\ W(t) < -x\}$, where we use the convention that $\inf\{t \in \varnothing\} = 2$. Note that both the τ's and ζ's are stopping tines. We call the largest k such that $\zeta_k \leqslant 1$ the number of upcrossings over the interval $[-x, x]$. Similarly, we define the number of downcrossing of the sequence $W(t)$ over the interval $[-x, x]$ by the number of upcrossings of the sequence $-W(t)$ over the interval $[-x, x]$.

Let K be the number of crossings (both up- and down-) of the Wiener process $\{W(t), 0 \leqslant t \leqslant 1\}$ over the interval $[-x, x]$.

Then

$$\begin{aligned}
&P\Big\{\sup_{0\leqslant t\leqslant 1}|W(t)| < x\Big\}\\
=&P\{W(1) \in [-x,x]\} - P\Big\{\sup_{0\leqslant t\leqslant 1}|W(t)| \geqslant x;\ W(1) \in [-x,x]\Big\}\\
=&P\{W(1)\in[-x,x]\}-2P\Big\{\sup_{0\leqslant t\leqslant 1}|W(t)|\geqslant x;\ W(\tau_0)=x,\ W(1) \in [-x,x]\Big\}\\
=&P\{W(1) \in [-x,x]\} - 2\sum_{k=0}^{\infty} P\Big\{W(\tau_0) = x,\ K = k,\ W(1) \in [-x,x]\Big\}\\
=&P\{W(1)\in[-x,x]\}-2\sum_{k=0}^{\infty}(-1)^k P\Big\{W(\tau_0)=x,\ K\geqslant k,\ W(1)\in[-x,x]\Big\}.
\end{aligned}$$

Reflecting the process $W(t)$ about x at $\zeta_{[(k+2)/2]}, \cdots, \zeta_1$ and reflecting the process about $-x$ at the $\tau_{[k/2]}, \cdots, \tau_1$, by reflection principle, we have

$$P\Big\{W(\tau_0)=x,\ K\geqslant k,\ W(1)\in[-x,x]\Big\}=P(W(1)\in[(2k+1)x, (2k+3)x]).$$

Then, the asserted identity is proved by noticing

$$P(W(1) \in [(2k+1)x, (2k+3)x]) = P(W(1) \in [-(2k+1)x, -(2k+3)x]).$$

Define

$$h(t) = \begin{cases} 1, & \text{if } \ 0 < t < x, \\ -1, & \text{if } \ x < t < 2x; \end{cases}$$

$$h(t) = h(-t); \qquad h(t) = h(t+x).$$

By the Fourier expansion we have

$$h(t) = \frac{4}{\pi} \sum_{k=0}^{\infty} \frac{(-1)^k}{2k+1} \cos\left(\frac{2k+1}{2x}\pi t\right).$$

By what being proved, we have

$$\begin{aligned} P\Big\{ \sup_{0 \leqslant t \leqslant 1} |W(t)| < x \Big\} &= \frac{1}{\sqrt{2\pi}} \sum_{k=-\infty}^{\infty} (-1)^k \int_{(2k-1)x}^{(2k+1)x} \mathrm{e}^{-t^2/2} \mathrm{d}t \\ &= \frac{1}{\sqrt{2\pi}} \int_{-\infty}^{\infty} h(t) \mathrm{e}^{-t^2/2} \mathrm{d}t \\ &= \frac{4}{\pi} \sum_{k=0}^{\infty} \frac{(-1)^k}{2k+1} \frac{1}{\sqrt{2\pi}} \int_{-\infty}^{\infty} \mathrm{e}^{-t^2/2} \cos\left(\frac{(2k+1)\pi}{2x} t\right) \mathrm{d}t \\ &= \frac{4}{\pi} \sum_{k=0}^{\infty} \frac{(-1)^k}{2k+1} \exp\left\{ - \frac{(2k+1)^2 \pi^2}{8x^2} \right\}, \end{aligned}$$

where we have used the fact that

$$\frac{1}{\sqrt{2\pi}} \int_{-\infty}^{\infty} \mathrm{e}^{-t^2/2} \cos \alpha t \mathrm{d}t = \mathrm{e}^{-\alpha^2/2}.$$

Choose only the term $k = 0$, we get the RHS inequality and take the three terms $k = -1, 0, 1$ we obtain the LHS inequality.

Remark. The series expansion of $P\left\{\sup_{0 \leqslant t \leqslant 1} |W(t)| < x\right\}$ in the above proof is itself a very important conclusion.

12.1.c (Csörgő and Révész). For any $0 < \varepsilon < 1$, there exists a constant $C = C(\varepsilon) > 0$ such that for any $0 < a < T$ and $x > 0$,

$$P\Big\{ \sup_{0 \leqslant t \leqslant T-a} \sup_{0 \leqslant s \leqslant a} |W(t+s) - W(t)| \geqslant x a^{1/2} \Big\} \leqslant C T a^{-1} \exp\{-x^2/(2+\varepsilon)\}.$$

Proof. For given $\varepsilon > 0$, define $\Delta = 1/\sqrt{1+\varepsilon/2}$ and $M = \frac{2\Delta}{1-\Delta}$. Select points $t_i = ia/M^2$, $i = 0, 1, \cdots, K-1$ and $t_K = T - a$ from the interval $[0, T-a]$ such that $0 < t_K - t_{K-1} < a/M^2$. Thus, we have $K \leqslant 1 + M^2T/a$.

Noticing $\Delta(1 + 2/M) = 1$, we have

$$P\Big\{\sup_{0\leqslant t\leqslant T-a}\sup_{0\leqslant s\leqslant a}|W(t+s)-W(t)| \geqslant xa^{1/2}\Big\} \leqslant \sum_{i=1}^{K}\Big(I_{i1}+I_{i2}+I_{i3}\Big), \tag{88}$$

where

$$\begin{aligned} I_{i1} &= P\Big\{\sup_{t_i\leqslant t\leqslant t_{i+1}}\sup_{0\leqslant s\leqslant a}|W(t+s)-W(t_i+s)| \geqslant \Delta M^{-1}xa^{1/2}\Big\} \\ &= P\Big\{\sup_{0<t\leqslant a/M^2}|W(t)| \geqslant \Delta M^{-1}xa^{1/2}\Big\} \\ &= 2P(|W(1)| \geqslant \Delta x) \leqslant Ce^{-x^2/(2+\varepsilon)}, \end{aligned}$$

where the first equality follows from the fact that $W(t+s) - W(t_i+s)$ has the same distribution as $W(t - t_i)$, and the second equality follows from 12.1.b. Similarly, we have

$$\begin{aligned} I_{i2} &= P\Big\{\sup_{t_i\leqslant t\leqslant t_{i+1}}|W(t)-W(t_i)| \geqslant \Delta M^{-1}xa^{1/2}\Big\} \\ &= P\Big\{\sup_{0<t\leqslant a/M^2}|W(t)| \geqslant \Delta M^{-1}xa^{1/2}\Big\} \\ &\leqslant Ce^{-x^2/(2+\varepsilon)}. \end{aligned}$$

Since $W(t_i+s)-W(t_i)$ has the same distribution as $W(s)$, by 12.1.b, we obtain

$$\begin{aligned} I_{i3} &= P\Big\{\sup_{0\leqslant s\leqslant a}|W(t_i+s)-W(t_i)| \geqslant \Delta xa^{1/2}\Big\} \\ &= P\Big\{\sup_{0<s\leqslant a}|W(s)| \geqslant \Delta xa^{1/2}\Big\} \\ &\leqslant \ Ce^{-x^2/(2+\varepsilon)}. \end{aligned}$$

The desired inequality from the above estimates.

12.2 Probability Estimate of Supremum of a Poisson Process

Let $\psi(t) = 2h(t+1)/t^2$ with $h(t) = t(\log t - 1) + 1$ for $t > 0$.

12.2.a. $P\{\sup_{0\leqslant t\leqslant b}(N(t)-t)^{\pm}/\sqrt{b}\geqslant x\}\leqslant\exp\left\{-\frac{x^2}{2}\psi\left(\frac{\pm x}{\sqrt{b}}\right)\right\}$

for any $x>0$ in the " + " case and for $0<x\leqslant\sqrt{b}$ in the " − " case.

Proof. For any $r>0$, $\{\exp(\pm r(N(t)-t)), 0\leqslant t\leqslant b\}$ are both submartingales. Then by a continuous parameter version of the Doob inequality 6.5.a,

$$
\begin{aligned}
&P\{\sup_{0\leqslant t\leqslant b}(N(t)-t)^{\pm}\geqslant x\}\\
&=\inf_{r>0}P\{\sup_{0\leqslant t\leqslant b}\exp(\pm r(N(t)-t))\geqslant\exp(rx)\}\\
&\leqslant\inf_{r>0}\exp(-rx)E\exp(\pm r(N(b)-b))\\
&\leqslant\inf_{r>0}\exp\{-rx+b(e^{\pm r}-1)\mp rb\}\\
&=\begin{cases}\exp\{x-(b+x)\log((b+x)/b)\} & \text{in the " + " case}\\ \exp\{-x+(b-x)\log(b/(b-x))\} & \text{in the " − " case}\end{cases}\\
&=\exp\left\{-(x^2/2b)\psi\left(\frac{\pm x}{b}\right)\right\}.
\end{aligned}
$$

Here the minimum is obtained by differentiating the exponent and solving. Now replace x by $x\sqrt{b}$ to get the desired inequality.

12.2.b. Let $q(t)$ be nondecreasing and $q(t)/\sqrt{t}$ be nonincreasing for $t\geqslant 0$ and let $0\leqslant a\leqslant(1-\delta)b<b\leqslant\delta<1$. Then for $x>0$,

$$P\{\sup_{a\leqslant t\leqslant b}(N(t)-t)^{\pm}/q(t)\geqslant x\}\leqslant\frac{3}{\delta}\int_a^b\frac{1}{t}\exp\left\{-(1-\delta)\gamma^{\pm}\frac{x^2}{2}\frac{q^2(t)}{t}\right\}\mathrm{d}t,$$

where $\gamma^-=1$ and $\gamma^+=\psi(xq(a)/a)$.

Proof. Let $A_n^{\pm}=\{\sup_{a\leqslant t\leqslant b}(N(t)-t)^{\pm}/q(t)\geqslant x\}$. Define $\theta=1-\delta$ and integers $0\leqslant J\leqslant K$ by

$$\theta^K<a\leqslant\theta^{K-1}\quad\text{and}\quad\theta^J<b\leqslant\theta^{J-1}\qquad(\text{we let }K=\infty\text{ if }a=0).$$

(from here on, θ^i denotes θ^i for $J\leqslant i<K$, but θ^K denotes a and θ^{J-1} denotes b. Note that (new θ^{i-1}) $\leqslant$(new θ^i)$/\theta$ is true for all $J\leqslant i\leqslant K$.)

Since q is nondecreasing, we have

$$\begin{aligned}P(A_n^{\pm}) &\leqslant P\{\max_{J\leqslant i\leqslant K}\sup_{\theta^i\leqslant t\leqslant\theta^{i-1}}(N(t)-t)^{\pm}/q(t)\geqslant x\}\\&\leqslant P\{\max_{J\leqslant i\leqslant K}\sup_{\theta^i\leqslant t\leqslant\theta^{i-1}}(N(t)-t)^{\pm}/q(\theta^i)\geqslant x\}\\&\leqslant\sum_{i=J}^{K}P\{\sup_{0\leqslant t\leqslant\theta^{i-1}}(N(t)-t)^{\pm}/\geqslant xq(\theta^i)\}.\end{aligned}$$

Consider A_n^- first. Similarly to the proof in 12.2.a, we have

$$\begin{aligned}P(A_n^-) &\leqslant\sum_{i=J}^{K}\exp\left\{-\frac{x^2q^2(\theta^i)}{2\theta^{i-1}}\right\}\\&\leqslant\sum_{i=J+1}^{K-1}\frac{1}{1-\theta}\int_{\theta^i}^{\theta^{i-1}}\frac{1}{t}\exp\left\{-\frac{x^2q^2(t)}{2t}\theta\right\}\mathrm{d}t\\&\quad+\exp\left\{-\frac{x^2q^2(a)}{2a}\theta\right\}+\exp\left\{-\frac{x^2q^2(\theta b)}{2b}\right\}\\&\leqslant\frac{3}{\delta}\int_a^b\frac{1}{t}\exp\left\{-(1-\delta)\frac{x^2q^2(t)}{2t}\right\}\mathrm{d}t.\end{aligned}$$

For A_n^+, using 12.2.a, as in case A_n^- we have

$$\begin{aligned}P(A_n^+) &\leqslant\sum_{i=J}^{K}\exp\left\{-\frac{x^2q^2(\theta^i)}{2\theta^{i-1}}\psi\left(\frac{xq(\theta^i)}{\theta^{i-1}}\right)\right\}\\&\leqslant\sum_{i=J}^{K}\exp\left\{-\frac{x^2q^2(\theta^i)}{2\theta^{i-1}}\psi\left(\frac{xq(a)}{a}\right)\right\}\\&\leqslant\frac{3}{\delta}\int_a^b\frac{1}{t}\exp\left\{-\frac{x^2q^2(t)}{2t}\theta\psi\left(\frac{xq(a)}{a}\right)\right\}\mathrm{d}t,\end{aligned}$$

as desired.

12.3 Fernique Inequality

Let d be a positive integer. $\mathscr{D}=\{t:t=(t_1,\cdots,t_d),\ a_j\leqslant t_j\leqslant b_j,\ j=1,\cdots,d\}$ with the usual Euclidean norm $\|\cdot\|$. Let $\{X(t),t\in\mathscr{D}\}$ be a centered Gaussian process satisfying that $0 < \Gamma^2 \equiv \sup_{t\in\mathscr{D}} EX(t)^2 < \infty$ and

$$E(X(t) - X(s))^2 \leqslant \varphi(\|t - s\|),$$

where $\varphi(\cdot)$ is a nondecreasing continuous function such that $\int_0^\infty \varphi(\mathrm{e}^{-y^2})$ $\mathrm{d}y < \infty$. Then for $\lambda > 0$, $x \geqslant 1$ and $A > \sqrt{2d\log 2}$ we have

$$P\left\{\sup_{t\in\mathscr{D}} X(t) \geqslant x\left\{\Gamma + 2(\sqrt{2}+1)A\int_1^\infty \varphi(\sqrt{d}\lambda 2^{-y^2})\mathrm{d}y\right\}\right\}$$

$$\leqslant (2^d + B)\left(\prod_{j=1}^d \left(\frac{b_j - a_j}{\lambda} + \frac{1}{2}\right)\right)\mathrm{e}^{-x^2/2},$$

where $B = \sum_{n=1}^\infty \exp\{-2^{n-1}(A^2 - 2d\log 2)\}$.

Proof. Put $\varepsilon_n = \lambda 2^{-2^n}$, $n = 0, 1, \cdots$. For $k = (k_1, \cdots, k_d)$ with $k_i = 0, 1, \cdots$, $k_{in} \equiv [(b_i - a_i)/\varepsilon_n]$, $i = 1, \cdots, d$, define $t_k^{(n)} = (t_{1k_1}^{(n)}, \cdots, t_{dk_d}^{(n)})$ in $\mathscr{D}$, where

$$t_{ik_i}^{(n)} = a_i + k_i\varepsilon_n, \quad i = 1, \cdots, d.$$

Let

$$T_n = \{t_k^{(n)}, k = 0, \cdots, k_n = (k_{1n}, \cdots, k_{dn})\},$$

which contains $N_n \equiv \prod_{i=1}^d k_{in}$ points, $N_n \leqslant \prod_{i=1}^d \{(2^{2^n}(b_i - a_i)/\lambda) + 1\}$. Then the set $\bigcup_{n=0}^\infty T_n$ is dense in $\mathscr{D}$ and $T_n \subset T_{n+1}$. For $j \geqslant 1$ let $x_j = xA\varphi(\sqrt{d}\varepsilon_{j-1})2^{j/2}$ and $g_j = 2^{(j-1)/2}$. Then

$$\begin{aligned}\sum_{j=1}^\infty x_j &= xA\sum_{j=1}^\infty \varphi(\sqrt{d}\lambda 2^{-2^{j-1}})2^{j/2} \\ &= xA\sum_{j=1}^\infty \varphi(\sqrt{d}\lambda 2^{-g_j^2})(2\sqrt{2}+2)(g_j - g_{j-1}) \\ &\leqslant 2(\sqrt{2}+1)xA\sum_{j=1}^\infty \int_{g_{j-1}}^{g_j} \varphi(\sqrt{d}\lambda 2^{-y^2})\mathrm{d}y \\ &\leqslant 2(\sqrt{2}+1)xA\int_1^\infty \varphi(\sqrt{d}\lambda 2^{-y^2})\mathrm{d}y.\end{aligned}$$

Therefore we have

$$P\left\{\sup_{t\in\mathscr{D}} X(t) \geqslant x\left(\Gamma + 2(\sqrt{2}+1)A\int_1^\infty \varphi(\sqrt{d}\lambda 2^{-y^2})\mathrm{d}y\right)\right\}$$

$$\leqslant P\left\{\sup_{n\geqslant 0}\sup_{t\in T_n} X(t) \geqslant x\Gamma + \sum_{j=1}^\infty x_j\right\}$$

$$= \lim_{n\to\infty} P\left\{\sup_{t\in T_n} X(t) \geqslant x\Gamma + \sum_{j=1}^n x_j\right\}.$$

Let

$$B_0 = \{\sup_{t\in T_0} X(t) \geqslant x\Gamma\}, \quad B_n = \left\{\sup_{t\in T_n} X(t) \geqslant \sum_{j=1}^n x_j\right\},$$

$$A_n = \left\{\sup_{t\in T_n} X(t) \geqslant x\Gamma + \sum_{j=1}^n x_j\right\}, \quad n \geqslant 1.$$

We have

$$\begin{aligned} P(A_n) &\leqslant P(B_{n-1}) + P(A_n B_{n-1}^c) \\ &\leqslant P(B_{n-1}) + P(B_n B_{n-1}^c) \\ &\leqslant P(B_0) + \sum_{j=1}^\infty P(B_j B_{j-1}^c), \end{aligned}$$

where

$$\begin{aligned} P(B_j B_{j-1}^c) &= P\left\{\bigcup_{t\in T_j}\left(X(t) \geqslant \sum_{k=1}^j x_k\right) \cap \bigcap_{s\in T_{j-1}}\left(X(s) < \sum_{k=1}^{j-1} x_k\right)\right\} \\ &\leqslant P\left\{\bigcup_{t\in T_j - T_{j-1}} \bigcup_{\substack{s\in T_{j-1} \\ \|t-s\|\leqslant\sqrt{d}\varepsilon_{j-1}}} (X(t) - X(s) \geqslant x_j)\right\} \\ &\leqslant \sum_{t\in T_j - T_{j-1}} \sum_{\substack{s\in T_{j-1} \\ \|t-s\|\leqslant\sqrt{d}\varepsilon_{j-1}}} P\{X(t) - X(s) \geqslant x_j\}. \end{aligned}$$

Noting the fact that there is only one point s in the set $\{s \in T_{j-1} : \|t-s\| \leqslant \sqrt{d}\varepsilon_{j-1}\}$ for any $t \in T_j - T_{j-1}$ and that

$$E(X(t) - X(s))^2 \leqslant \varphi^2(\|t-s\|) \leqslant \varphi^2(\sqrt{d}\varepsilon_{j-1}),$$

we have

$$
\begin{aligned}
&P(B_jB_{j-1}^c)\\
&\leqslant \sum_{t\in T_j-T_{j-1}} \sum_{\substack{s\in T_{j-1}\\ \|t-s\|\leqslant\sqrt{d}\varepsilon_{j-1}}} P\left\{N(0,1)\geqslant \frac{x_j}{\varphi(\sqrt{d}\varepsilon_{j-1})}\right\}\\
&\leqslant \prod_{i=1}^d \left(2^{2^j}\frac{b_i-a_i}{\lambda}+1\right) P\{N(0,1)\geqslant Ax2^{j/2}\}\\
&\leqslant 2^{2^j d}\prod_{i=1}^d \left(\frac{b_i-a_i}{\lambda}+\frac{1}{2}\right)\frac{1}{2\sqrt{\pi}}\mathrm{e}^{-A^2x^22^{j-1}}\\
&=\frac{1}{2\sqrt{\pi}}\prod_{i=1}^d \left(\frac{b_i-a_i}{\lambda}+\frac{1}{2}\right)\mathrm{e}^{2^j d\log 2-(2^{j-1}A^2-1/2)x^2}\mathrm{e}^{-x^2/2}\\
&<d^{-2^j((A^2/2)-d\log 2)}\prod_{i=1}^d \left(\frac{b_i-a_i}{\lambda}+\frac{1}{2}\right)\mathrm{e}^{-x^2/2}.
\end{aligned}
$$

Noting that $A>\sqrt{2d\log 2}$, we obtain

$$
\sum_{j=1}^{\infty} P(B_jB_{j-1}^c)\leqslant B\prod_{i=1}^d \left(\frac{b_i-a_i}{\lambda}\vee 1\right)\mathrm{e}^{-x^2/2}.
$$

On the other hand,

$$
\begin{aligned}
P(B_0)&=P\{\sup_{t\in T_0} X(t)\geqslant x\Gamma\}\\
&\leqslant 2^d\prod_{i=1}^d \left(\frac{b_i-a_i}{\lambda}+\frac{1}{2}\right)P(N(0,1)\geqslant x)\\
&<2^d\prod_{i=1}^d \left(\frac{b_i-a_i}{\lambda}+\frac{1}{2}\right)\mathrm{e}^{-x^2/2}.
\end{aligned}
$$

Hence

$$
P(A_n)\leqslant (2^d+B)\prod_{i=1}^d \left(\frac{b_i-a_i}{\lambda}+\frac{1}{2}\right)\mathrm{e}^{-x^2/2},
$$

which gives the desired inequality immediately.

12.4 Borell Inequality

Let $\{X(t),t\in T\}$ be a zero-mean-value separable Gaussian process, and the sample paths are bounded almost everywhere. Denote $\|X\|=$

$\sup_{t\in T}|X(t)|$. Then for any $x>0$,

$$P\{||X||-E||X||>x\}\leqslant 2\exp\{-x^2/(2\sigma_T^2)\},$$

where $\sigma_T^2=\sup_{t\in T}EX(t)^2$.

Remark. The Gaussian process in the inequality can be replaced by a Gaussian variable in Banach space. And $E||X||$ can also be replaced by the median of $||X||$. If so, then the Borell inequality can be yielded by the equi-perimeter inequality 12.6 (see Ledoux and Talagrand (1991)). When $T=[0,h]$, we have the following accurate big-bias conclusion.

12.5 Tail Probability of Gaussian Process

Let $\{X(t),t\in T\}$ is a zero-mean-value separable Gaussian process, $EX^2(t)=1,t\geqslant 0$,

$$\Gamma(s,t)\equiv \mathrm{Cov}(X(s),X(t))=1-C_0|s-t|^\alpha+o(|s-t|^\alpha),\quad |s-t|\to 0,$$

where $0<\alpha\leqslant 2, C_0>0$. Then for any $h>0$,

$$\lim_{x\to\infty}\frac{P\{\max_{t\in[0,h]}X(t)>x\}}{x^{2/\alpha}(1-\Phi(x))}=hC_0^{1/\alpha}H_\alpha,$$

in which $H_\alpha=\lim_{T\to\infty}\int_0^\infty \mathrm{e}^s P\{\sup_{0\leqslant t\leqslant T}Y(t)>s\}\,\mathrm{d}s/T>0$, $Y(t)$ is a Gaussian process, which has mean-value $EY(t)=-|t|^\alpha$ and covariance function $\mathrm{Cov}(Y(s), Y(t))=-|s-t|^\alpha+|s|^\alpha+|t|^\alpha$.

Remark. This conclusion was obtained by Pickands (1969a,b), and Qualls and Watanabe (1972) generalized it to the case of R^k. It can be proved that $H_1=1$, $H_2=1/\sqrt{\pi}$. Shao (1996) presented the estimations of the upper and lower bounds of H_α.

12.6 Tail Probability of Randomly Signed Independent Processes

For $-\infty<a<b<\infty$, put $T=[a,b]$ or $[a,\infty)$. D denotes the set of all functions of T that are right continuous and possess left-hand limits at each point. $\mathscr{D}$ denotes the σ-algebra generated by the finite dimensional subsets of D. Let $\{X_i(t),t\in T\}$, $i=1,\cdots,n$, be independent processes on $(D, \mathscr{D})$ that are independent of the iid. Rademacher r.v.'s $\varepsilon_1, \cdots, \varepsilon_n$. Then for any $x > 0$,

$$P\Big\{\max_{1\leqslant k\leqslant n}\sup_{t\in T}\Big|\sum_{j=1}^{k}\varepsilon_j X_j(t)\Big| > x\Big\} \leqslant 2P\Big\{\sup_{t\in T}\Big|\sum_{j=1}^{n}\varepsilon_j X_j(t)\Big| > x\Big\}.$$

Proof. Without loss of generality assume that $a = 0$, $0 < b \leqslant \infty$. Let $S_0(t) \equiv 0$, $S_k(t) = \sum_{j=1}^{k} \varepsilon_j X_j(t)$ and $K_m = \{j/2^m : 0 \leqslant j \leqslant 2^m b\}$. Denote

$$A_k = \{\max_{0\leqslant j<k}\sup_{t\in T} S_j^+(t) \leqslant x < \sup_{t\in T} S_k^+(t)\},$$

$$A_{km} = \{\max_{0\leqslant j<k}\sup_{t\in T} S_j^+(t) \leqslant x < \sup_{t\in K_m} S_k^+(t)\},$$

$$J_{km} = \min\{j : S_k(j/2^m) > x\}, \qquad k = 1, \cdots, n.$$

Furthermore, let $K = \bigcup_{m=1}^{\infty} K_m$, which is countably dense in T. Note that $\sup_{t\in K_m}|f(t)| \to \sup_{t\in T}|f(t)|$ as $m \to \infty$ for all $f \in D$. Whereas $\sup_{t\in T}|f(t)|$ may not equal $f(\tau)$ for some $\tau \in T$ (it could equal some $f(\tau_-)$), it is the case that $\sup_{t\in K_m}|f(t)|$ equals $f(\tau)$ for some $\tau \in K_m$. Now noting the symmetry of S_k we have

$$\begin{aligned}
P\{\sup_{t\in T} S_n^+(t) > x\} &\geqslant \sum_{k=1}^{n} P\Big\{A_k \bigcap (\sup_{t\in T} S_n^+(t) > x)\Big\} \\
&= \sum_{k=1}^{n} \lim_{m\to\infty} P\Big\{A_{km} \bigcap (\sup_{t\in K_m} S_n^+(t) > x)\Big\} \\
&= \sum_{k=1}^{n} \lim_{m\to\infty} \sum_{j=0}^{b2^m} P\Big\{A_{km} \bigcap (\sup_{t\in K_m} S_n^+(t) > x) \\
&\qquad \bigcap (J_{km} = j)\Big\} \\
&\geqslant \sum_{k=1}^{n} \lim_{m\to\infty} \sum_{j=0}^{b2^m} P\Big\{A_{km} \bigcap (S_n(j/2^m) \geqslant S_k(j/2^m)) \\
&\qquad \bigcap (J_{km} = j)\Big\} \\
&= \sum_{k=1}^{n} \lim_{m\to\infty} \sum_{j=0}^{b2^m} P\Big\{A_{km} \bigcap (J_{km} = j)\Big\} \\
&\qquad P\{S_n(j/2^m) - S_k(j/2^m) \geqslant 0\}
\end{aligned}$$

$$\geqslant \frac{1}{2}\sum_{k=1}^{n} \lim_{m\to\infty} \sum_{j=0}^{b2^m} P\Big\{A_{km} \bigcap (J_{km} = j)\Big\}$$
$$= \frac{1}{2}\sum_{k=1}^{n} \lim_{m\to\infty} P(A_{km}) = \frac{1}{2}\sum_{k=1}^{n} P(A_k)$$
$$= \frac{1}{2} P\{\max_{1\leqslant k\leqslant n} \sup_{t\in T} S_k^+(t) > x\}.$$

For S_n^- we have the same conclusion. The inequality is proved.

Let $(\Omega, \mathscr{F}, P)$ be a probability space, $\mathscr{F} = \{\mathscr{F}_t \subset \mathscr{F} : t \geqslant 0\}$ be a family of σ-algebras satisfying $\mathscr{F}_s \subset \mathscr{F}_t$ for $s < t$. A process $X = \{X(t), t \geqslant 0\}$ is $\mathscr{F}$-adapted, if $X(t)$ is $\mathscr{F}_t$-measurable for all $t \geqslant 0$. A process X is predictable if it is measurable with respect to the σ-algebra on $(0, \infty) \times \Omega$ generated by the collection of adapted processes which are left continuous on $(0, \infty)$. A process X has a property locally if there exists a localizing sequence of stopping times $\{T_k : k \geqslant 1\}$ such that $T_k \to \infty$ a.s. as $k \to \infty$ and the process $X(\cdot \wedge T_k)$ has the desired property for each $k \geqslant 1$. An extension of the definition of a (sub- or super-)martingale (see Chapter 6) in the continuous parameter case is immediate. The following inequality is a generalization of Doob's inequality 6.5.a.

12.7 Tail Probability of Adaptive Process

Suppose that X is an adapted nonnegative process with right-continuous sample paths, that Y is an adapted process with nondecreasing right-continuous sample paths and $Y(0) = 0$ a.s. and that X is dominated by Y, i.e., $EX(T) \leqslant EY(T)$ for all stopping times T.

(i) If T is predictable, then for any $x > 0$, $y > 0$ and all stopping times T,

$$P\Big\{\sup_{0\leqslant t\leqslant T} |X(t)| \geqslant x\Big\} \leqslant \frac{1}{x} E(Y(T) \wedge y) + P\{Y(T) \geqslant y\}.$$

(ii) If $\sup_{t>0} |Y(t) - Y(t-)| < a$, then for any $x > 0$, $y > 0$ and all stopping times T,

$$P\Big\{\sup_{0\leqslant t\leqslant T} |X(t)| \geqslant x\Big\} \leqslant \frac{1}{x} E(Y(T) \wedge (y + a)) + P\{Y(T) \geqslant y\}.$$

Proof. We first show that

$$P\Big\{\sup_{0\leqslant t\leqslant T}|X(t)|\geqslant x\Big\}\leqslant\frac{1}{x}EY(T). \tag{89}$$

Let $S=\inf\{s\leqslant T\wedge n: X(s)\geqslant x\}$, $T\wedge n$ if the set is empty. Thus S is a stopping time and $S\leqslant T\wedge n$. Hence

$$\begin{aligned}EY(T)&\geqslant EY(S)\geqslant EX(S)\\&\geqslant EX(S)I\Big(\sup_{0\leqslant t\leqslant T\wedge n}X(t)\geqslant x\Big)\\&\geqslant xP\Big\{\sup_{0\leqslant t\leqslant T\wedge n}X(t)\geqslant x\Big\}.\end{aligned}$$

Letting $n\to\infty$ yields (89).

Denote $X_t^*=\sup_{0\leqslant s\leqslant t}X(s)$. To prove (i), we shall show that for $x>0$, $y>0$ and all predictable stopping times S,

$$P\{X_{S-}^*\geqslant x\}\leqslant\frac{1}{x}E\{Y(S-)\wedge y\}+P\{Y(S-)\geqslant y\}. \tag{90}$$

Then (i) follows from (90) applied to the processes $X^T\equiv X(\cdot\wedge T)$ and $Y^T\equiv Y(\cdot\wedge T)$ with the predictable stopping time $S\equiv\infty$.

Let $R=\inf\{t: Y(t)\geqslant y\}$. Then $R>0$ by the right continuity of Y and is predictable since Y is predictable. Thus $R\wedge S$ is predictable and with a sequence S_n of stopping times satisfying $S_n<R\wedge S$, $S_n\to R\wedge S$ and $\{X_{(R\wedge S)-}^*\geqslant x\}\subset\liminf_{n\to\infty}\{X_{S_n}^*\geqslant x-\varepsilon\}$,

$$\begin{aligned}&P\{X_{S-}^*\geqslant x\}=P\{Y_{S-}<y,X_{S-}^*\geqslant x\}+P\{Y_{S-}\geqslant y,X_{S-}^*\geqslant x\}\\&\leqslant P\{I(Y_{S-}<y)X_{S-}^*\geqslant x\}+P\{Y_{S-}\geqslant y\}\\&\leqslant P\{X_{(R\wedge S)-}^*\geqslant x\}+P\{Y_{S-}\geqslant y\}\\&\leqslant\liminf_{n\to\infty}P\{X_{S_n}^*\geqslant x-\varepsilon\}+P\{Y_{S-}\geqslant y\}\\&\leqslant\frac{1}{x-\varepsilon}\liminf_{n\to\infty}EY_{S_n}+P\{Y_{S-}\geqslant y\}\\&=\frac{1}{x-\varepsilon}EY_{(R\wedge S)-}+P\{Y_{S-}\geqslant y\}\\&\leqslant\frac{1}{x-\varepsilon}E(Y_{S-}\wedge y)+P\{Y_{S-}\geqslant y\}.\end{aligned}$$

Letting $\varepsilon\downarrow 0$ yields (90) which in turn implies (i). The argument for (ii) is similar.

See Lenglart (1977).

12.8 Tail Probability on Submartingale

Let $(|S_t|, \mathscr{F}_t)$, $0 \leqslant t \leqslant b$ be a submartingale whose sample paths are right (or left) continuous. Suppose $S(0) = 0$ and $\nu(t) \equiv ES^2(t) < \infty$ on $[0, b]$. Let $q > 0$ be an increasing right (or left) continuous function on $[0, b]$. Then

$$P\left\{\sup_{0\leqslant t\leqslant b} |S(t)|/q(t) \geqslant 1\right\} \leqslant \int_0^b (q(t))^{-2}\mathrm{d}\nu(t).$$

Proof. By the right (left) continuity of the sample paths and $S(0) = 0$, using 6.6.c, we have

$$\begin{aligned}
&P\left\{\sup_{0\leqslant t\leqslant b} |S(t)|/q(t) \leqslant 1\right\}\\
=&P\left\{\max_{0\leqslant j\leqslant 2^n} |S(bj/2^n)|/q(bj/2^n) \leqslant 1 \text{ for all } n \geqslant 1\right\}\\
=&\lim_{n\to\infty} P\left\{\max_{0\leqslant j\leqslant 2^n} |S(bj/2^n)|/q(bj/2^n) \leqslant 1\right\}\\
\geqslant&\lim_{n\to\infty}\left\{1 - \sum_{j=1}^{2^n}(E(S^2(bj/2^n) - S^2(b(j-1)/2^n))/q^2(bj/2^n))\right\}\\
=&1 - \lim_{n\to\infty}\sum_{j=1}^{2^n}\frac{1}{q^2(bj/2^n)}\{\nu(bj/2^n) - \nu(b(j-1)/2^n)\}\\
=&1 - \int_0^b (q(t))^{-2}\mathrm{d}\nu(t),
\end{aligned}$$

where the convergence comes from the monotone convergence theorem. In fact, we only need S to be separable and q to be increasing.

We shall denote the canonical Gaussian measure on R^N by γ_N, which is the probability measure on R^N with density

$$(2\pi)^{-N/2}\exp(-|x|^2/2).$$

$\Phi(x)$ denotes also the d.f. of $N(0,1)$ on R^1. S^{N-1} denotes the Euclidean unit sphere in R^N equipped with its geodesic distance ρ and normalized Haar measure σ_{N-1}.

Let B be a Banach space equipped with a norm $\|\cdot\|$ such that, for some countable subset D of the unit ball of B', the dual space of all continuous linear functionals, $\|x\| = \sup_{f\in D}|f(x)|$ for all x in B. We say that X is a Gaussian r.v. in B if $f(X)$ is measurable for every f in D and if every finite linear combination $\sum_i \alpha_i f_i(X)$, $\alpha_i \in R^1$, $f_i \in D$ is Gaussian. Let $M = M(X)$ be a median of $\|X\|$; $Ef^2(X)$, $f \in D$, are called weak variances. A sequence $\{X_n, n \geqslant 1\}$ of r.v.'s with values in B is called a symmetric sequence if, for every choice of signs ± 1, $\{\pm X_n, n \geqslant 1\}$ has the same distribution as $\{X_n, n \geqslant 1\}$ (i.e. for each n, $(\pm X_1, \cdots, \pm X_n)$ has the same distribution as $(X_1, \cdots, X_n)$ in B^n).

Most of the inequalities for real r.v.'s such as symmetrization inequalities, Lévy's inequality, Jensen's inequality, Ottaviani's inequality, Hoffmann-Jϕrgensen's inequality, Khintchine's inequality etc., can be extended to the Banach valued r.v.'s case.

See Birnbaum and Marshall (1961).

12.9 Tail Probability of Independent Sum in B-Space

Let $X_1, \cdots, X_n$ be independent B-valued r.v.'s with $E\|X_j\|^p < \infty$ for some $p > 2$, $j = 1, \cdots, n$. Then there exists a constant $C = C(p)$ such that for any $t > 0$,

$$P\left\{\left\|\sum_{j=1}^n X_j\right\| \geqslant t + 37p^2 E\left\|\sum_{j=1}^n X_j\right\|\right\}$$
$$\leqslant 16\exp\{-t^2/144\Lambda_n\} + C\sum_{j=1}^n \|X_j\|^p/t^p,$$

where $\Lambda_n = \sup_{|f|\leqslant 1}\left\{\sum_{j=1}^n Ef^2(X_j)\right\}$.

The proof can be found in Einmahl (1993).

We state the following important inequality without the proof. The details may be found, for example, in Ledoux and Talagrand (1991).

12.10 Isoperimetric Inequalities

12.10.a (on a sphere). Let A be a Borel set in S^{N-1} and H be a cap (i.e. a ball for the geodesic distance ρ) with the same measure $\sigma_{N-1}(H) = \sigma_{N-1}(A)$. Then

$$\sigma_{N-1}(A_r) \geqslant \sigma_{N-1}(H_r),$$

where $A_r = \{x \in S^{N-1} : \rho(x, A) < r\}$ is the neighborhood of order r of A for the geodesic distance. In particular, if $\sigma_{N-1}(A) \geqslant 1/2$ (and $N \geqslant 3$), then

$$\sigma_{N-1}(A_r) \geqslant 1 - \left(\frac{\pi}{8}\right)^{1/2} \exp\{-(N-2)r^2/2\}.$$

12.10.b (in a Gaussian space). Let A be a Borel set in R^N and H be a half-space $\{x \in R^N; < x, u >< \lambda\}$, $u \in R^N$, $\lambda \in [-\infty, \infty]$, with the same Gaussian measure $\gamma_N(H) = \gamma_N(A)$. Then, for any $r > 0$, $\gamma_N(A_r) \geqslant \gamma_N(H_r)$ where, accordingly, A_r is the Euclidean neighborhood of order r of A. Equivalently,

$$\Phi^{-1}(\gamma_N(A_r)) \geqslant \Phi^{-1}(\gamma_N(A)) + r$$

and in particular, if $\gamma_N(A) \geqslant 1/2$,

$$1 - \gamma_N(A_r) \leqslant 1 - \Phi(r) \leqslant \frac{1}{2} \exp(-r^2/2).$$

Remark. Denoting the positive integer set by Z, we can generalize 12.10.b to measure $\gamma = \gamma_\infty$ in the infinite-dimension space R^Z, which is the product of one-dimension standard Gaussian distribution. By 12.10.b and cylindrical approximation, we have

$$\Phi^{-1}(\gamma_*(A_r)) \geqslant \Phi^{-1}(\gamma(A)) + r, \tag{91}$$

where γ_* is the inner measure and A is a Borel set in R^Z, $r > 0$, A_r is the Hilbert neighborhood of order r of A, i.e., $A_r = A + rB_2 = \{x = a + rh : a \in A, h \in R^N, |h| \leqslant 1\}$ where B_2 is the unit ball of l_2, the space of all real sequences $x = \{x_n\}$ for which $\|x\|_2 = \left(\sum_{n=1}^{\infty} x_n^2\right)^{1/2} < \infty$.

12.11 Ehrhard Inequality

For any convex set A in R^n, Borel set B and any $0 \leqslant \lambda \leqslant 1$,

$$\Phi^{-1}(\gamma_N(\lambda A + (1-\lambda)B)) \geqslant \lambda\Phi^{-1}(\gamma_N(A)) + (1-\lambda)\Phi^{-1}(\gamma_N(B)),$$

where $\lambda A + (1-\lambda)B = \{\lambda a + (1-\lambda)b : a \in A, b \in B\}$.

Refer to the proof in Latala (1996).

Remark. When A and B are both convex sets, the Ehrhard inequality can be called Brunn-Minkowski type inequality. 12.10.b can be yielded from 12.11.

12.12 Tail Probability of Normal Variable in B-Space

Let X be a Gaussian r.v. in B with median $M = M(X)$ and supremum of weak variances $\sigma^2 = \sigma^2(X)$. Then, for any $t > 0$,

$$P\{|\|X\| - M| > t\} \leqslant 2(1 - \Phi(t/\sigma)) \leqslant \exp(-t^2/2\sigma^2)$$

and

$$P\{\|X\| > t\} \leqslant 4\exp\{-t^2/8E\|X\|^2\}.$$

Proof. Let $A = \{x \in R^Z : \|x\| \leqslant M\}$. Then $\gamma(A) \geqslant 1/2$. By (91) we have $\gamma_*(A_t) \geqslant \Phi(t)$. Now, if $x \in A_t$, then $x = a + th$ where $a \in A$ and $|h| \leqslant 1$. Noting that

$$\sigma = \sigma(X) = \sup_{f\in D}(Ef^2(X))^{1/2} = \sup_{|h|\leqslant 1}\|h\|,$$

we have

$$\|x\| \leqslant M + t\|h\| \leqslant M + t\sigma$$

and therefore $A_t \subset \{x : \|x\| \leqslant M + \sigma t\}$. Applying the same argument to $A = \{x : \|x\| \geqslant M\}$ clearly concludes the proof of the first inequality. From it, and from $\sigma^2 \leqslant E\|X\|^2$, $M^2 \leqslant 2E\|X\|^2$, we obtain the second one.

12.13 Gaussian Measure on Symmetric Convex Sets

Let μ be a centered Gaussian measure in separable Banach space E, X and Y be two zero-mean-value Gaussian random elements in E. Then for any $0 < \lambda < 1$, and any two symmetric convex sets A and B in E,

$$\mu(A\bigcap B) \geqslant \mu(\lambda A)\mu((1-\lambda^2)^{1/2}B),$$

$$P\{X \in A, Y \in B\} \geqslant P\{X \in \lambda A\}P\{Y \in (1-\lambda^2)^{1/2}B\}.$$

Proof. We might as well assume that $E = R^n$. Let (X', Y') be the independent copy of (X, Y), and denote $a = (1-\lambda^2)^{1/2}/\lambda$. It is easy to see that $X - aX'$ and $Y + Y'/a$ do not intersect, and are thus independent of each other. With $X - aX'$ and X/λ identically distributed, and so is $Y + Y'/a$ and $Y/\sqrt{1-\lambda^2}$. By the Anderson inequality 2.3.b, we obtain

$$\begin{aligned} P\{X \in A, Y \in B\} &\geqslant P\{(X,Y) + (-aX', Y'/a) \in A \times B\} \\ &= P\{X - aX' \in A, Y + Y'/a \in B\} \\ &= P\{X - aX' \in A\}P\{Y + Y'/a \in B\} \\ &= P\{X \in \lambda A\}P\{Y \in \sqrt{1-\lambda^2}B\}. \end{aligned}$$

See Li (1999).

12.14 Equivalence of Moments of B-Gaussian Variables

Let X be a Gaussian r.v. in B. Then all the moments of $\|X\|$ are equivalent (and equivalent to $M = M(X)$) in the sense that for any $0 < p, q < \infty$, there exists a constant K_{pq} depending only on p and q such that

$$(E\|X\|^p)^{1/p} \leqslant K_{pq}(E\|X\|^q)^{1/q}.$$

Proof. Integrating the first inequality of 12.12 we obtain

$$\begin{aligned} E|\|X\| - M|^p &= \int_0^\infty P\{|\|X\| - M| > t\}\mathrm{d}t^p \\ &\leqslant \int_0^\infty \exp(-t^2/2\sigma^2)\mathrm{d}t^p \leqslant (K\sqrt{p}\sigma)^p \end{aligned}$$

for some numerical constant K. Now, this inequality is stronger than what we need since $\sigma \leqslant 2M$ and M can be majorized by $(2E\|X\|^q)^{1/q}$ for every $q > 0$. The proof is complete.

12.15 Contraction Principle

12.15.a. Let $f : R_+ \to R_+$ be convex, $\{\varepsilon_n, n \geqslant 1\}$ be a Rademacher sequence. For any finite sequence $\{x_n\}$ in a Banach space B and any real numbers $\{\alpha_n\}$ such that $|\alpha_n| \leqslant 1$ for each n, we have

$$Ef\left(\left\|\sum_n \alpha_n\varepsilon_n x_n\right\|\right) \leqslant Ef\left(\left\|\sum_n \varepsilon_n x_n\right\|\right).$$

Furthermore, for any $t > 0$,

$$P\left\{\left\|\sum_n \alpha_n\varepsilon_n x_n\right\| > t\right\} \leqslant 2P\left\{\left\|\sum_n \varepsilon_n x_n\right\| > t\right\}.$$

Proof. The function

$$(\alpha_1, \cdots, \alpha_N) \to Ef\left(\left\|\sum_{n=1}^N \alpha_n\varepsilon_n x_n\right\|\right)$$

is convex. Therefore, on the compact convex set $[-1, 1]^N$, it attains its maximum at an extreme point, that is a point $(\alpha_1, \cdots, \alpha_N)$ such that

$\alpha_n = \pm 1$. For such values of α_n, by symmetry, both terms in the first inequality are equal. This proves this inequality.

Concerning the second one, replacing α_n by $|\alpha_n|$, we may assume by symmetry that $\alpha_n \geqslant 0$. Moreover, by identical distribution, we suppose that $\alpha_1 \geqslant \cdots \geqslant \alpha_N \geqslant \alpha_{N+1} = 0$. Put $S_n = \sum_{j=1}^{n} \varepsilon_j x_j$. Then

$$\sum_{j=1}^{N} \alpha_j \varepsilon_j x_j = \sum_{n=1}^{N} \alpha_n (S_n - S_{n-1}) = \sum_{n=1}^{N} (\alpha_n - \alpha_{n+1}) S_n.$$

It follows that

$$\left\| \sum_{j=1}^{N} \alpha_j \varepsilon_j x_j \right\| \leqslant \max_{1 \leqslant n \leqslant N} \|S_n\|.$$

We conclude the proof by a version of Lévy's inequality 5.4 in the Banach space case.

12.15.b. Let $f : R_+ \to R_+$ be convex. Let $\{\eta_n\}$ and $\{\xi_n\}$ be two symmetric sequences of real r.v.'s such that for some constant $K \geqslant 1$ and every n and $t > 0$,

$$P\{|\eta_n| > t\} \leqslant KP\{|\xi_n| > t\}.$$

Then, for any finite sequence $\{x_n\}$ in a Banach space,

$$Ef\left(\left\| \sum_n \eta_n x_n \right\|\right) \leqslant Ef\left(K \left\| \sum_n \xi_n x_n \right\|\right).$$

Proof. Let $\{\delta_n\}$ be independent of $\{\eta_n\}$ such that $P\{\delta_n = 1\} = 1 - P\{\delta_n = 0\} = 1/K$ for each n. Then, for any $t > 0$,

$$P\{|\delta_n \eta_n| > t\} \leqslant P\{|\xi_n| > t\}.$$

Taking inverses of the d.f.'s, it is easily seen that the sequences $\{\delta_n \eta_n\}$ and $\{\xi_n\}$ can be constructed on some rich enough probability space in such a way that

$$|\delta_n \eta_n| \leqslant |\xi_n| \quad \text{a.s.} \quad \text{for} \quad \text{each} \quad n.$$

From the contraction principle 12.15.a and the symmetry assumption, it follows that

$$Ef\left(\left\| \sum_n \delta_n \eta_n x_n \right\|\right) \leqslant Ef\left(\left\| \sum_n \xi_n x_n \right\|\right).$$

The proof is then completed via Jensen's inequality 8.4.a applied to the sequence $\{\delta_n\}$ since $E\delta_n = 1/K$.

12.16 Symmetrization Inequalities in B-Space

Let $f : R_+ \to R_+$ be convex, then for any finite sequence $\{X_n\}$ of independent mean zero (i.e. $Eg(X_n) = 0$ for all $g \in D$) r.v.'s in B such that $Ef(\|X_n\|) < \infty$ for each n,

$$Ef\left(\frac{1}{2}\left\|\sum_n \varepsilon_n X_n\right\|\right) \leqslant Ef\left(\left\|\sum_n X_n\right\|\right) \leqslant Ef\left(2\left\|\sum_n \varepsilon_n X_n\right\|\right),$$

where $\{\varepsilon_n\}$ is a Rademacher sequence which is independent of $\{X_n\}$.
Proof. Let $\{X_n'\}$ be an independent copy of the sequence $\{X_n\}$ and be also independent of $\{\varepsilon_n\}$. Put $X_n^s = X_n - X_n'$. Then, by Fubini's theorem, Jensen's inequality, zero mean and convexity, recalling the Remark of 8.6, we have

$$\begin{aligned} Ef\left(\left\|\sum_n X_n\right\|\right) &\leqslant Ef\left(\left\|\sum_n X_n^s\right\|\right) = Ef\left(\left\|\sum_n \varepsilon_n X_n^s\right\|\right) \\ &\leqslant Ef\left(2\left\|\sum_n \varepsilon_n X_n\right\|\right). \end{aligned}$$

Conversely, by the same argument,

$$\begin{aligned} Ef\left(\frac{1}{2}\left\|\sum_n \varepsilon_n X_n\right\|\right) &\leqslant Ef\left(\frac{1}{2}\left\|\sum_n \varepsilon_n X_n^s\right\|\right) \\ &= Ef\left(\frac{1}{2}\left\|\sum_n X_n^s\right\|\right) \leqslant Ef\left(\left\|\sum_n X_n\right\|\right). \end{aligned}$$

The inequality is proved.

12.17 Decoupling Inequality

Let $\{X_n, n \geqslant 1\}$ be a sequence of real independent r.v.'s, let $\{X_{ln}, n \geqslant 1\}$ be independent copies of $\{X_n, n \geqslant 1\}$ for $1 \leqslant l \leqslant k$. Furthermore, let $f_{i_1,\cdots,i_k}$ be elements of a Banach space such that $f_{i_1,\cdots,i_k} = 0$ unless the $i_1,\cdots,i_k$ are distinct. Then for any $1 \leqslant p \leqslant \infty$, we have

$$\left\|\sum_{i_1,\cdots,i_k} f_{i_1,\cdots,i_k} X_{i_1}\cdots X_{i_k}\right\|_p \leqslant (2k+1)^k \left\|\sum_{i_1,\cdots,i_k} f_{i_1,\cdots,i_k} X_{1i_1}\cdots X_{ki_k}\right\|_p,$$

where $\|\xi\|_p = (E\|\xi\|^p)^{1/p}$ for a Banach space valued r.v. ξ.

Proof. Let $m_n = EX_n$, $\bar{X}_n = X_n - m_n$, $\bar{X}_{ln} = X_{ln} - m_n$, $l = 1, \cdots, k$, and let $X = \{X_n, n \geqslant 1\}$, $X_l = \{X_{ln}, n \geqslant 1\}$, $l = 1, \cdots, k$, and $\mathscr{X}_j = (X_1, \cdots, X_j)$. At first we show that for $1 \leqslant r \leqslant k$,

$$\left\| \sum_{i_1,\cdots,i_r} f_{i_1,\cdots,i_r} \bar{X}_{1i_1} \cdots \bar{X}_{ri_r} \right\|_p \leqslant 2^r \left\| \sum_{i_1,\cdots,i_r} f_{i_1,\cdots,i_r} X_{1i_1} \cdots X_{ri_r} \right\|_p . \tag{92}$$

Indeed, by interchangeability,

$$\begin{aligned}
&\left\| \sum_{i_1,\cdots,i_r} f_{i_1,\cdots,i_r} \bar{X}_{1i_1} \cdots \bar{X}_{ri_r} \right\|_p \\
=&\left\| \sum_{i_1,\cdots,i_r} f_{i_1,\cdots,i_r} (X_{1i_1} - m_{i_1}) \cdots (X_{ri_r} - m_{i_r}) \right\|_p \\
=&\left\| \sum_{(\delta_1,\cdots,\delta_r)\in\{0,1\}^r} \sum_{i_1,\cdots,i_r} f_{i_1,\cdots,i_r} X_{1i_1}^{\delta_1} \cdots X_{ri_r}^{\delta_r} m_{i_1}^{1-\delta_1} \cdots m_{i_r}^{1-\delta_r} \right\|_p \\
\leqslant&\sum_{j=0}^{r} \binom{r}{j} \left\| \sum_{i_1,\cdots,i_r} f_{i_1,\cdots,i_r} X_{1i_1} \cdots X_{ji_j} m_{i_{j+1}} \cdots m_{i_r} \right\|_p \\
=&\sum_{j=0}^{r} \binom{r}{j} \left\| \sum_{i_1,\cdots,i_r} f_{i_1,\cdots,i_r} X_{1i_1} \cdots X_{ji_j} E(X_{j+1,i_{j+1}} | \mathscr{X}_j) \cdots E(X_{r,i_r} | \mathscr{X}_j) \right\|_p \\
=&\sum_{j=0}^{r} \binom{r}{j} \left\| E\left(\sum_{i_1,\cdots,i_r} f_{i_1,\cdots,i_r} X_{1i_1} \cdots X_{ji_j} X_{j+1,i_{j+1}} \cdots X_{r,i_r} | \mathscr{X}_j \right) \right\|_p \\
\leqslant&\sum_{j=0}^{r} \binom{r}{j} \left\| \sum_{i_1,\cdots,i_r} f_{i_1,\cdots,i_r} X_{1i_1} \cdots X_{ji_j} X_{j+1,i_{j+1}} \cdots X_{ri_r} \right\|_p \\
=&2^r \left\| \sum_{i_1,\cdots,i_r} f_{i_1,\cdots,i_r} X_{1i_1} \cdots X_{ri_r} \right\|_p .
\end{aligned}$$

Here Jensen's inequality was used.

Similarly, we have

$$\begin{aligned}
&\left\| \sum_{i_1,\cdots,i_k} f_{i_1,\cdots,i_k} X_{i_1} \cdots X_{i_k} \right\|_p \\
\leqslant&\sum_{r=0}^{k} \binom{k}{r} \left\| \sum_{i_1,\cdots,i_k} f_{i_1,\cdots,i_k} \bar{X}_{i_1} \cdots \bar{X}_{i_r} m_{i_{r+1}} \cdots m_{i_k} \right\|_p
\end{aligned}$$

$$=\sum_{r=0}^{k}\binom{k}{r}\bigg\|E\bigg(\sum_{i_1,\cdots,i_k} f_{i_1,\cdots,i_k}(\bar{X}_{1i_1}+\cdots+\bar{X}_{ri_1})\cdots$$

$$(\bar{X}_{1i_r}+\cdots+\bar{X}_{ri_r})m_{i_{r+1}}\cdots m_{i_k}|\mathscr{X}_1\bigg)\bigg\|_p$$

$$\leqslant\sum_{r=0}^{k}\binom{k}{r}\bigg\|\sum_{i_1,\cdots,i_k} f_{i_1,\cdots,i_k}(\bar{X}_{1i_1}+\cdots+\bar{X}_{ri_1})\cdots$$

$$(\bar{X}_{1i_r}+\cdots+\bar{X}_{ri_r})m_{i_{r+1}}\cdots m_{i_k}\bigg\|_p.$$

Put $\mathscr{G}_r=\sigma\left(\sum_{j=1}^{r}X_j\right)$. The last expression is equal to

$$\sum_{r=0}^{k}\binom{k}{r}\bigg\|r^rE\bigg(\sum_{i_1,\cdots,i_k} f_{i_1,\cdots,i_k}\bar{X}_{1i_1}\cdots\bar{X}_{ri_r}m_{i_{r+1}}\cdots m_{i_k}|\mathscr{G}_r\bigg)\bigg\|_p$$

$$\leqslant\sum_{r=0}^{k}\binom{k}{r}\bigg\|r^r\sum_{i_1,\cdots,i_k} f_{i_1,\cdots,i_k}\bar{X}_{1i_1}\cdots\bar{X}_{ri_r}m_{i_{r+1}}\cdots m_{i_k}\bigg\|_p$$

$$\leqslant\sum_{r=0}^{k}\binom{k}{r}(2r)^r\bigg\|\sum_{i_1,\cdots,i_k} f_{i_1,\cdots,i_k}X_{1i_1}\cdots X_{ri_r}m_{i_{r+1}}\cdots m_{i_k}\bigg\|_p$$

$$=\sum_{r=0}^{k}\binom{k}{r}(2r)^r\|E\bigg(\sum_{i_1,\cdots,i_k} f_{i_1,\cdots,i_k}X_{1i_1}\cdots X_{ri_r}X_{r+1,i_{r+1}}\cdots X_{ki_k}|\mathscr{X}_r\bigg)\bigg\|_p$$

$$\leqslant(2k+1)^k\bigg\|\sum_{i_1,\cdots,i_k} f_{i_1,\cdots,i_k}X_{1i_1}\cdots X_{ki_k}\bigg\|_p,$$

as desired. Here the second inequality is due to (92).

References

Billingsley P. 1999. Convergence of Probability Measures. 2nd ed.. New York: J. Wiley.

Birnbaum Z, Marshall A. 1961. Some multivariate Chebyshev inequalities with extensions to continuous parameter processes. Ann. Math. Statist., 32: 687-703.

Choi Y K, Lin Z Y. 1998. A version of Fernique lemma for Gaussian processes. East Asian Math., 14: 99-106.

Csőrgö M, Révész P. 1981. Strong Approximations in Probability and Statistics. New York: Academic Press.

de la Peña V H, Montgomery-Smith S J, Szulga J. 1994. Contraction and decoupling inequalities for multilinear forms and U-statistics. Ann. Probab., 22: 1745-1765.

Einmahl U. 1993. Toward general law of the iterated logarithm in Banach space. Ann. Probab., 21: 2012-2045.

Fernique X. 1964. Continuite des processus Gaussiens. C. R. Acad. Sci. Paris, 258: 6058-6060.

Hill T P, Houdré C. 1999. Advances in Stochastic Inequalities. AMS special session on stochastic inequalities and their applications, 1997. Contem. Math., 234.

Jain N C, Marcus M B. 1974. Integrability of infinite sums of vector-valued random variables. Trans. Amer. Math. Soc., 212: 1-36.

Latala R. 1996. A note on the Ehrhard's inequality. Studia Math., 118: 169-174.

Ledoux M, Talagrand M. 1991. Probability in Banach Spaces. Berlin: Springer-Verlag.

Lenglart E. 1977. Relation de domination entre deux processus. Ann. Inst. Henri poincare, 13: 171-179.

Li W V. 1999. A Gaussian correlation inequality and its applications to small ball probabilities. Elect. Comm. Probab., 4: 111-118.

Li W V, Shao Q M. 2001. Gaussian processes: inequalities, small ball probabilities and applications//Edited by Rao C R and Shanbhag D. Stochastic Processes: Theory and Methods, Handbook of Statistics. Amsterdam, 19: 533-597.

Lin Z Y, Lu C L, Zhang L X. 2001. Path Properties of Gaussian Processes. Beijing: Science Press & Hangzhou: Zhejiang Univ. Press.

Mörters P, Peres Y. 2008. Brownian Motion. Manuscript.

Pickands J III. 1969a. Upcrossing probabilities for stationary Gaussian processes. Trans. Amer. Math. Soc., 145: 51-73.

Pickands J III. 1969b. Asymptotic probabilities of the maximum in a stationary Gaussian processes. Trans. Amer. Math. Soc., 145: 75-86.

Qualls C, Watanable H. 1972. Asymptotic properties of Gaussian Processes. Ann. Math. Statist., 43: 580-596.

Shao Q M. 1996. Bounds and estimators of a basic constant in extreme value theory of Gaussian processes. Statist. Sinica, 6: 245-257.

Tong Y L. 1984. Inequalities in Statistics and Probability: Proceedings of the Symposium on Inequalities in Statistics and Probability, 1982. Lincoln. IMS.

Zeitfracht Medien GmbH
Ferdinand-Jühlke-Straße 7
99095 Erfurt, Deutschland
produktsicherheit@kolibri360.de